METHODS IN MOLECULAR BIOLOGY™

Series Editor
John M. Walker
School of Life Sciences
University of Hertfordshire
Hatfield, Hertfordshire, AL10 9AB, UK

For further volumes:
http://www.springer.com/series/7651

Olfactory Receptors

Methods and Protocols

Edited by

Chiquito J. Crasto

Division of Research, Department of Genetics, University of Alabama at Birmingham, Birmingham, AL, USA

Humana Press

Editor
Chiquito J. Crasto
Division of Research
Department of Genetics
University of Alabama at Birmingham
Birmingham, AL, USA

ISSN 1064-3745 ISSN 1940-6029 (electronic)
ISBN 978-1-4939-6202-0 ISBN 978-1-62703-377-0 (eBook)
DOI 10.1007/978-1-62703-377-0
Springer New York Dordrecht Heidelberg London

Cover illustration Caption: The cover figure depicts a computational model of the human olfactory receptor hOR17-209 with the model of an isopentylacetate odorant ligand docked in the protein binding region. The OR-odorant system is embedded in a simulated plasma membrane. Water molecules surround the plasma membrane. Created by Peter C. Lai and Chiquito J. Crasto.

Printed on acid-free paper

Humana Press is a brand of Springer
Springer is part of Springer Science+Business Media (www.springer.com)

Dedication

For
Hemant
Jim
Ed
Jin-an
Gordon

Dedication

Preface

... Some natures are so sensitive to certain smells; and it would even be a very fine question to study both in its pathological and physiological relation ...

from *Madame Bovary*
–Gustave Flaubert

A paper by Linda Buck and Richard Axel in the journal, *Cell*, in 1999, informed us of the discovery of an olfactory receptor gene. This tremendous discovery was rewarded by the Nobel Prize for Medicine in 2004. The discovery of ORs spurred research in identifying olfactory receptors as well as other chemosensory receptors such as taste and pheromone receptors, for several species.

When mammalian genomes were published, olfactory receptors were identified as constituting the largest families in these genomes. It appears that the higher the order of the species (whose olfactory repertoire has been mined and assembled from its genome), arguably, the lower is the acuity of olfactory function. This is evidenced by the relative number of nonfunctioning genes in the gene families, the percentage of these being highest in humans. Such a large gene repertoire should then conceal within it a gold mine of information as to the evolution of olfaction from lower order to higher order species. It also hints to the variability of the sequence (or structure)–function dichotomy, which allows fewer (despite its being a super-family) genes to discriminate several thousand odors singly or as complex mixtures.

Despite the best efforts of several and despite landmark discoveries and experimental ingenuity, challenges in the pursuit of research related to ORs exist. These hinder the functional characterization of olfactory receptors. Deorphanizing ORs—identifying odorants that will elicit an excitatory response from an OR—is difficult. This is because of the costs involved in testing odorants for an OR's response, a predictive notion for matching ORs and odorants being unavailable. This difficulty is exacerbated because of the promiscuous nature of OR–odor interactions. Protein expression, as is typically attempted in heterologous systems, has also proved difficult. Olfactory receptor repertoires consisting of several hundred genes have to be cataloged and made available to the community, informing of aspects related to the genomics, proteomics, and functional assessments. ORs are membrane-bound proteins, for which there currently exists no experimentally determined structure. Their protein structures have to be determined using ab initio or semi-empirical methods.

The chapters contained in this volume, titled *Olfactory Receptors: Methods and Protocols*, were chosen for the depiction of research that has resulted in significant contributions to addressing the issues mentioned in the above paragraph. Indeed, the parts within this volume (i.e. knowledge dissemination of ORs, theoretical assessments of OR structure and function, development and use of expression systems and experimental functional analysis) are designed to meet the above. The primary content of each chapter is written in the format of a laboratory manual, in a style that is readable to scientists as well as aspiring scientists.

This is in keeping with the general format for all volumes in the series. The chapters also provide a support-basis for methodological development for those involved in OR, GPCR, as well as membrane protein research.

ORs belong to the family of GPCRs, whose ubiquitous functionality has long been recognized and is being increasingly appreciated. At the time of completion of the development of this volume came a momentous announcement. In the words of the Nobel Committee: "*Robert Lefkowitz and Brian Kobilka are awarded the 2012 Nobel Prize in Chemistry for groundbreaking discoveries that reveal the inner workings of an important family of such receptors: G-protein–coupled receptors.*" (http://www.nobelprize.org/nobel_prizes/chemistry/laureates/2012/). ORs, because of the size of the gene family, the large variability in sequence (sustained by a uniformly similar three-dimensional structure), and the consequences to OR function, become, in several respects, a good testing ground for GPCR function.

Consequences of odor perception articulated by the prescient Flaubert notwithstanding, there is evidence that olfactory dysfunction can point to neurological disorders. Odor perception is perhaps more complex than specific odor recognition. It is possible that every individual possesses a unique olfactory fingerprint that is an agglomeration of a personal olfactory repertoire, a unique combination of neuronal processes and the health of the olfactory processing regions. ORs are responsible for the first step—odorant capture—that catalyze the cascade of events that results in olfaction. Understanding the mechanism of OR function will go a long way towards contributing to our knowledge of olfaction.

Chiquito J. Crasto
Birmingham, AL, USA

Acknowledgements

To Professor John Walker, whose editorship of such a colossal body of work—closing in on a thousand volumes—is nothing short of stupendous. Judging from his attention to detail in helping bring the 18 chapters presented in this volume to print, it is easy to imagine that every chapter in every volume in this series bears his personal touch. To Mrs. Walker, who assuredly plays a key role in the administration of this series.

To Professor Gordon Shepherd, under whose tutelage I was exposed to olfactory receptor research and who continues to provide valuable counsel.

To Dr. Michael Singer and past members of the research group of Professor William Goddard, specifically Drs. Vaidehi Nagarajan and Wely Floriano, all of whose work serve as a knowledgebase for my research, the computational assessments of olfactory receptors.

To all the authors who have contributed to this volume, especially the corresponding authors: Eleonora Alfinito, Ricardo Araneda, Kaveh Baghaei, Yehuda Beh-Shahar, Richard Benton, Jerome Golebiowski, Kara Hoover, Dietmar Krautwurst, Charles Luetje, Joel Mainland, Bettina Malnic, Valery Matarazzo, Yoshihito Niimura, Tsviya Olender, Micherl Renou, Guenhaël Sanz, and Shuguang Zhang. This volume has been in development for some time and some of you have had to be very patient.

To the staff at Humana Press and Springer, both in the USA and in India, whose tireless efforts have been critical to making this volume a success. To members of the production team at Humana Press who helped me initiate the process of creation of this volume, to Anne Meagher, Patrick Marton and David Casey from the Springer, New York, and especially to Priya Ranganathan of SPi Global Technologies for her diligence, patience and forbearance in the final stages of the production of this volume, I owe a large debt of gratitude.

To my dad, who passed away around the time I was thinking of working on a volume on olfactory receptors, to my mother, my brother, and my sister who continue to inspire and sustain me.

And when all's said and done and the dust's settled, I am fortunate to come home to my Budu, my Munder and my Avver.

Contents

Contributors

ELEONORA ALFINITO • *Dipartimento di Ingegneria dell'Innovazione, Università del Salento, Lecce, Italy; Consorzio Nazionale Interuniversitario per le Scienze Fisiche della Materia (CNISM), Rome, Italy*

RICARDO C. ARANEDA • *Department of Biology, University of Maryland, College Park, MD, USA*

KAVEH A. BAGHAEI • *Cell Physiology Department, Ruhr-Bochum University of Germany, Bochum, Germany*

GAUTAM BAHL • *North State Radiology Medical Group Inc, Chico, CA, USA*

YEHUDA BEN-SHAHAR • *Department of Biology, Washington University, St. Louis, MO, USA; Pulmonary and Critical Care Medicine, Department of Medicine, Washington University School of Medicine, St. Louis, Missouri, USA.*

RICHARD BENTON • *Faculty of Biology and Medicine, Center for Integrative Genomics, University of Lausanne, Lausanne, Switzerland*

ANA CASTRO • *Department of Molecular and Cellular Pharmacology, Miller School of Medicine, University of Miami, Miami, FL, USA*

LANDRY CHARLIER • *Institut de Chimie de Nice, UMR CNRS, Université de Nice Sophia Antipolis, Nice, France*

BRIAN COOK • *Center for Bits and Atoms, Massachusetts Institute of Technology, Cambridge, MA, USA*

KAROLINA CORIN • *Center for Bits and Atoms, Massachusetts Institute of Technology, Cambridge, MA, USA*

CHIQUITO J. CRASTO • *Division of Research, Department of Genetics, University of Alabama at Birmingham, Birmingham, AL, USA*

CLAIRE A. DE MARCH • *Institut de Chimie de Nice, UMR CNRS, Université de Nice Sophia Antipolis, Nice, France*

JEROME GOLEBIOWSKI • *Institut de Chimie de Nice, UMR CNRS, Université de Nice Sophia Antipolis, Nice, France*

XIAOLING GU • *Department of Biology, Washington University, St. Louis, MO, USA*

KARA C. HOOVER • *Anthropology Department, University of Alaska Fairbanks, Fairbanks, AK, USA*

LORRA HYLAND • *School of Nursing, University of Alabama at Birmingham, Birmingham, AL, USA*

NICHOLAS KATSANIS • *Duke University, Durham, NC, USA*

MATTHIAS KOTTHOFF • *Fraunhofer-Institute for Molecular Biology and Applied Ecology Department Environmental and Food Analysis Auf dem Aberg, Schmallenberg, Germany*

DIETMAR KRAUTWURST • *German Research Center for Food Chemistry, Leibniz Institute, Freising, Germany*

PETER C. LAI • *Division of Research, Department of Genetics, University of Alabama at Birmingham, Birmingham, AL, USA*

DORON LANCET • *Department of Molecular Genetics, Weizmann Institute of Science, Rehovot, Israel; The Crown Human Genome Center, Weizmann Institute of Science, Rehovot, Israel*
PHILIPPE LUCAS • *INRA, UMR1272 PISC, Versailles, France*
CHARLES W. LUETJE • *Department of Molecular and Cellular Pharmacology, Miller School of Medicine, University of Miami, Miami, FL, USA*
JOEL D. MAINLAND • *Monell Chemical Senses Center, Philadelphia, PA, USA*
BETTINA MALNIC • *Department of Biochemistry, Institute of Chemistry, University of São Paulo, São Paulo, Brazil*
LUIS N. MARENCO • *Center for Medical Informatics, Yale University School of Medicine, New Haven, CT, USA*
VALÉRY MATARAZZO • *Centre de Recherche en Neurobiologie et Neurophysiologie de Marseille, CRN2M, CNRS UMR6231, Université Paul Cézanne, Marseille, France*
HIROAKI MATSUNAMI • *Duke University, Durham, NC, USA*
PERRY L. MILLER • *Center for Medical Informatics, Yale University School of Medicine, New Haven, CT, USA*
NOAM NATIV • *Department of Molecular Genetics, Weizmann Institute of Science, Rehovot, Israel; The Crown Human Genome Center, Weizmann Institute of Science, Rehovot, Israel*
ANDREW S. NICHOLS • *Department of Molecular and Cellular Pharmacology, Miller School of Medicine, University of Miami, Miami, FL, USA; Booz Allen Hamilton Inc., Baltimore, MD, USA*
YOSHIHITO NIIMURA • *Department of Bioinformatics, Medical Research Institute, Tokyo Medical and Dental University, Tokyo, Japan*
TSVIYA OLENDER • *Department of Molecular Genetics, Weizmann Institute of Science, Rehovot, Israel; The Crown Human Genome Center, Weizmann Institute of Science, Rehovot, Israel*
EDITH PAJOT-AUGY • *Unité de Neurobiologie de l'Olfaction et Modélisation en Imagerie & Equipe Biologie de l'Olfaction et Biosenseurs, INRA, Jouy-en-Josas, France*
ZITA PETERLIN • *Department of Biology, Columbia University, New York, NY, USA*
JEREMY POUSSET • *Dipartimento di Matematica e Fisica, Ennio de Giorgi, Università del Salento, Lecce, Italy*
LINO REGGIANI • *Dipartimento di Matematica e Fisica, Ennio de Giorgi, Università del Salento, Lecce, Italy; Consorzio Nazionale Interuniversitario per le Scienze Fisiche della Materia (CNISM), Rome, Italy*
MICHEL RENOU • *INRA, UMR1272 PISC, Versailles, France*
CATHERINE RONIN • *Insitut de Neurobiologie Intégrative et Adaptative, UMR6149 CNRS et Université de Provence, Marseille, France*
MICHAEL SAINA • *Faculty of Biology and Medicine, Center for Integrative Genomics, University of Lausanne, Lausanne, Switzerland*
GUENHAËL SANZ • *Unité de Neurobiologie de l'Olfaction et Modélisation en Imagerie & Equipe Biologie de l'Olfaction et Biosenseurs, INRA, Jouy-en-Josas, France*
GORDON M. SHEPHERD • *Department of Neurobiology, Yale University School of Medicine, New Haven, CT, USA*
BENJAMIN L. SHERMAN • *Department of Molecular and Cellular Pharmacology, Miller School of Medicine, University of Miami, Miami, FL, USA*

JING SHI • *School of Public Health, Emory University, Atlanta, GA, USA*
RICHARD S. SMITH • *Department of Biology, University of Maryland, College Park, MD, USA*
JÉRÉMIE TOPIN • *Institut de Chimie de Nice, UMR CNRS, Université de Nice Sophia Antipolis, Nice, France*
RIXIN WANG • *Center for Medical Informatics, Yale University School of Medicine, New Haven, CT, USA*
JASON R. WILLER • *Duke University, Durham, NC, USA*
SHUGUANG ZHANG • *Center for Bits and Atoms, Massachusetts Institute of Technology, Cambridge, MA, USA*

Part I

Knowledge Dissemination of Olfactory Receptors

Chapter 1

Databases in SenseLab for the Genomics, Proteomics, and Function of Olfactory Receptors

Luis N. Marenco, Gautam Bahl, Lorra Hyland, Jing Shi, Rixin Wang, Peter C. Lai, Perry L. Miller, Gordon M. Shepherd, and Chiquito J. Crasto

Abstract

We present here, the salient aspects of three databases: Olfactory Receptor Database (ORDB) is a repository of genomics and proteomics information of ORs; OdorDB stores information related to odorous compounds, specifically identifying those that have been shown to interact with olfactory rectors; and OdorModelDB disseminates information related to computational models of olfactory receptors (ORs). The data stored among these databases is integrated. Presented in this chapter are descriptions of these resources, which are part of the SenseLab suite of databases, a discussion of the computational infrastructure that enhances the efficacy of information storage, retrieval, dissemination, and automated data population from external sources.

Key words Databases, Olfactory receptors, Computational modeling, Odorants, Informatics, Data integration

1 Introduction

Research into olfactory receptors (ORs) is a relatively recent field, about two decades old (1). OR function results from the interaction of an OR with an odorant—the first in a series of steps leading to olfaction (2, 3). Two major events provided the impetus for research into ORs and other chemosensory receptors (CRs): (1) The Nobel Prize for Medicine awarded in 2004 for the discovery ORs (4) and (2) the recognition of OR genes as constituting the largest gene families in the human and most mammalian genomes (5–9). Olfactory Receptor Database (ORDB), which stores information related to these OR genes and described in this chapter, (10–16) has been at the forefront of information dissemination related to ORs for more than 15 years.

Electro-olfactograms (17–20), and calcium imaging (21–25) are two commonly used modes of experimentally determining OR excitation. The central question that these methodologies attempt

Chiquito J. Crasto (ed.), *Olfactory Receptors: Methods and Protocols*, Methods in Molecular Biology, vol. 1003, DOI 10.1007/978-1-62703-377-0_1, © Springer Science+Business Media, LLC 2013

to address is the nature of the interaction between an OR and odor molecule ligand. The first evidence for OR response to odor-binding came from an experimental study by Breer et al. (26) and a computational study by Singer & Shepherd (27). This was followed by a computational analysis by Singer (28, 29) of experimental results by Zhao et al. (30); the latter study showed a preference of the I7 receptor for the octanal odorant. The closeness of the fit of computational model to experiment supported the notion of the existence of a binding pocket within which odor molecules interact with receptor amino acid residues. The first combinatorial study of odor-mediated excitation of different identified ORs was published in 1999 for 14 mouse ORs and several odorants—mostly carboxylic acids and their derivatives. Later, electro-olfactograms were used to determine responses of the rat OR I7 to a rational panel of 90 odors, which were eventually narrowed to ten octanal molecules that differed by substitutions, branches, and unsaturation (31).

Other functional activation studies have shown that phenylic aldehydes and ketones activate mouse OR M71 and OR912-93—both extensively characterize (32). Luu et al. showed that key receptor residues are responsible for interacting with odor ligands in zebra-fish and goldfish (33). Touhara and coworker identified ORs that could be excited by lyral, eugenol, vanillin, and ethyl vanillin (34–42). Matarazzo and coworkers have used baculovirus constructs to perform functional analysis of two human ORs (22). Shirokova et al. established that the type of G-protein in the transduction cascade plays an important role in determining whether an odor will be an agonist or an antagonist (21). Schmiedeberg showed that the interactions between the amino acid residues in an OR and an odor are mediated not only by electrostatic interactions but supported by van der Waals forces (43). These are a sampling of earlier functional analysis for ORs. More recently, two combinatorial studies have been published for several ORs with odors with a wide range of functional groups. While most ORs are narrowly tuned to agonists for not more than one feature, e.g., functional groups, functional characterization of the mouse ORs MOR256-3 (44) and MOR256-31 (45) shows that these ORs respond to odors possessing a wide range of functional groups. The promiscuous nature of the interactions—one OR has been shown to bind more than one odorant, which in turn will bind more than one OR—adds to the challenge of obtaining mechanistic insights into OR-odor interactions.

Functional characterization of ORs involve determining which odorants excite, inhibit or exhibit no impact on an OR. These studies pinpoint specific amino acid residues within the OR binding region that are responsible for odorant binding.

There is currently no experimentally derived structure of ORs. An OR is a GPCR—a G(TP-binding) Protein Coupled Receptor.

Structurally, GPCRs are known to be composed of seven transmembrane helices, which traduce the plasma membrane. These helices are connected by three intracellular and three extra-cellular loops and possess an extracellular N-terminus and an intracellular C-terminus (46). There is some evidence that GPCRs and ORs with anomalous structures might exist (47). Computational experiments have been carried out to determine the structure of ORs. Like most GPCRs, ORs are often modeled (28, 29, 48–53) on the structural template of rhodopsin—whose X-ray crystallographically derived structure has been solved with improving resolution (54, 55). Recently, more GPCRs have been structurally elucidated (56–59).

2 Materials

The SenseLab suite of databases uses a multi-tier application architecture composed of the following:

1. Interface front end modules written in the C# language.
2. EAV/CR business-data-logic layer written in the VB.NET language.
3. Back-end data storage layer resides on an Oracle 11g database server.
4. SenseLab uses Internet Information Server as a Web server hosted on a Windows 2008 R2 server.
5. Some data-integration and automated population software is written in PERL (Practical Extraction and Report Language).

3 Methods

3.1 ORDB

1. ORDB is a repository of information related primarily to ORs. Other chemosensory receptors are also represented in ORDB. They are *C. elegans* chemosensory receptors, fungal pheromone receptors, insect ORs, taste receptors, and vomeronasal receptors. The total number of receptors stored in ORDB currently numbers close to 14,500.
2. Each olfactory (or chemosensory receptor) is described by attributes. Each attribute is associated with one or more value, depending on the attribute. Figure 1 represents an OR with ORDB name ORL11. The number "11" associated with "ORL" merely indicates the order in which this gene was entered into the database, and bears no special significance. One of the attributes "Common Name" identifies the receptor by the name given to it by the research group where the OR was first cloned or identified.

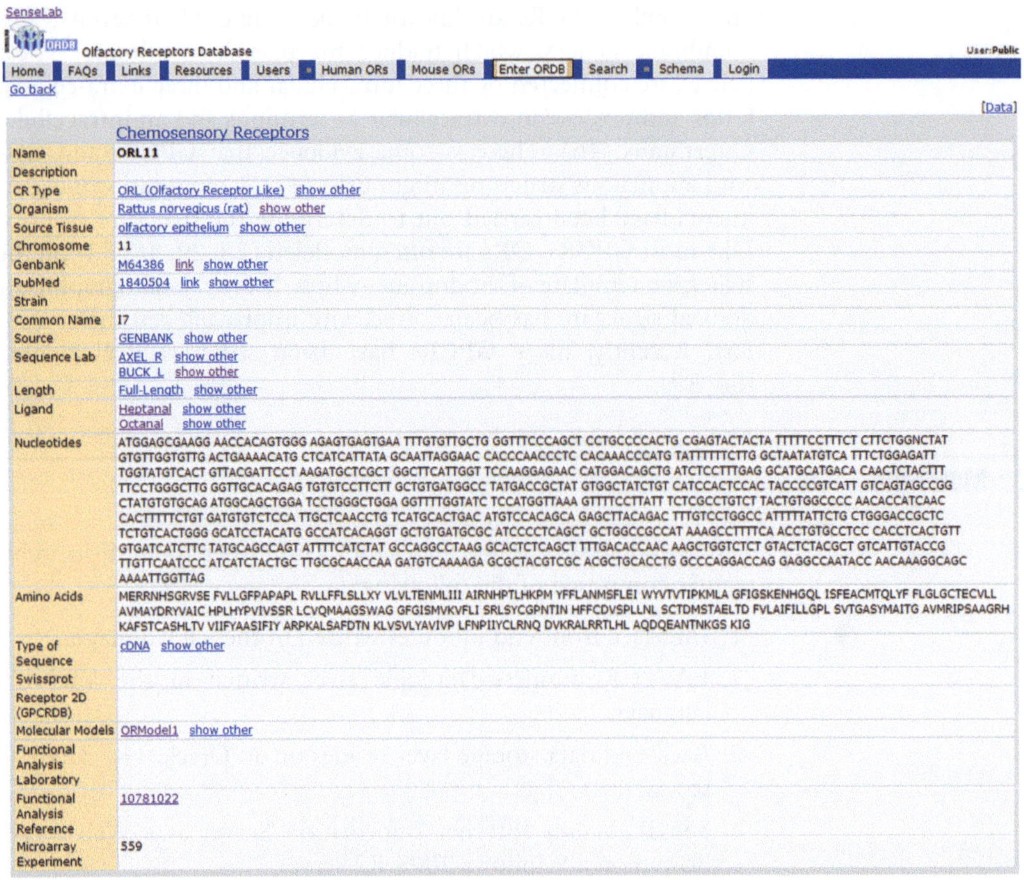
SenseLab

ORDB Olfactory Receptors Database User:Public

Home | FAQs | Links | Resources | Users | Human ORs | Mouse ORs | Enter ORDB | Search | Schema | Login

Go back

[Data]

	Chemosensory Receptors
Name	ORL11
Description	
CR Type	ORL (Olfactory Receptor Like) show other
Organism	Rattus norvegicus (rat) show other
Source Tissue	olfactory epithelium show other
Chromosome	11
Genbank	M64386 link show other
PubMed	1840504 link show other
Strain	
Common Name	I7
Source	GENBANK show other
Sequence Lab	AXEL R show other BUCK L show other
Length	Full-Length show other
Ligand	Heptanal show other Octanal show other
Nucleotides	ATGGAGCGAAGG AACCACAGTGGG AGAGTGAGTGAA TTTGTGTTGCTG GGTTTCCCAGCT CCTGCCCCACTG CGAGTACTACTA TTTTTCCTTTCT CTTCTGGNCTAT GTGTTGGTGTTG ACTGAAAACATG CTCATCATTATA GCAATTAGGAAC CACCCAACCCTC CACAAACCCATG TATTTTTTCTTG GCTAATATGTCA TTTCTGGAGATT TGGTATGTCACT GTTACGATTCCT AAGATGCTCGCT GGCTTCATTGGT TCCAAGGAGAAC CATGGACAGCTG ATCTCCTTTGAG GCATGCATGACA CAACTCTACTTT TTCCTGGGCTTG GGTTGCACAGAG TGTGTCCTTCTT GCTGTGATGGCC TATGACCGCTAT GTGGCTATCTGT CATCCACTCCAC TACCCCGTCATT GTCAGTAGCCGG CTATGTGTGCAG ATGGCAGCTGGA TCCTGGGCTGGA GGTTTTGGTATC TCCATGGTTAAA GTTTTCCTTATT TCTCGCCTGTCT TACTGTGGCCCC AACACCATCAAC CACTTTTTCTGT GATGTGTCTCCA TTGCTCAACCTG TCATGCACTGAC ATGTCCACAGCA GAGCTTACAGAC TTTGTCCTGGCC ATTTTTATTCTG CTGGGACCGCTC TCTGTCACTGGG GCATCCTACATG GCCATCACAGGT GCTGTGATGCGC ATCCCCTCAGCT GCTGGCCGCCAT AAAGCCTTTTCA ACCTGTGCCTCC CACCTCACTGTT GTGATCATCTTC TATGCAGCCAGT ATTTTCATCTAT GCCAGGCCTAAG GCACTCTCAGCT TTTGACACCAAC AAGCTGGTCTCT GTACTCTACGCT GTCATTGTACCG TTGTTCAATCCC ATCATCTACTGC TTGCGCAACCAA GATGTCAAAAGA GCGCTACGTCGC ACGCTGCACCTG GCCCAGGACCAG GAGGCCAATACC AACAAAGGCAGC AAAATTGGTTAG
Amino Acids	MERRNHSGRVSE FVLLGFPAPAPL RVLLFFLSLLXY VLVLTENMLIII AIRNHPTLHKPM YFFLANMSFLEI WYVTVTIPKMLA GFIGSKENHGQL ISFEACMTQLYF FLGLGCTECVLL AVMAYDRYVAIC HPLHYPVIVSSR LCVQMAAGSWAG GFGISMVKVFLI SRLSYCGPNTIN HFFCDVSPLLNL SCTDMSTAELTD FVLAIFILLGPL SVTGASYMAITG AVMRIPSAAGRH KAFSTCASHLTV VIIFYAASIFIY ARPKALSAFDTN KLVSVLYAVIVP LFNPIIYCLRNQ DVKRALRRTLHL AQDQEANTNKGS KIG
Type of Sequence	cDNA show other
Swissprot	
Receptor 2D (GPCRDB)	
Molecular Models	ORModel1 show other
Functional Analysis Laboratory	
Functional Analysis Reference	10781022
Microarray Experiment	559

Fig. 1 The ORDB entry for the rat I7 olfactory receptor. The ORDB name for this entry is ORL11. The figure illustrates the attributes on the *left hand side* of the page and their values on the *right hand side*. Values that are stored as independent classes are presented as links

3. The first attribute for this receptor is "Name" for which the value is "ORL11." The CR (Chemosensory Receptor) type (since receptors other than ORs are deposited in ORDB) type is ORL or OR like. The name of the organism (rat), the source tissue for the gene (olfactory epithelium), and chromosome number (11) provide additional information about the receptor.
4. The links to the article(s) that described this gene in PubMed and the GenBank entry which served as the source of this receptor are automatically generated when this information is entered into ORDB. The links allow the user to access the abstract for the article or the source for the gene.
5. The amino acid and nucleotide sequences for this gene are available.

6. The source of the information (most often GenBank) for this receptor is indicated. Other sources for this receptor are sequences supplied by researchers themselves, sometimes prior to publication (ORDB offers a service to researchers, who can obtain a login ID and use the database to store sequences privately (see Notes 1–3)).
7. The values for the "Sequence Laboratory" attribute contain links with researchers' names in whose laboratories the receptor gene was first studied. Clicking on the links "Axel_RA" and "Buck_LB" will take users to Web pages dedicated to information related to the laboratories for Drs. Richard Axel and Linda Buck, respectively. ORDB endeavors to be an information hub of the chemosensory receptor community. The page for each laboratory contains links to contact information—telephone numbers, e-mail, and laboratory address for these researchers.
8. The "Length" of the sequence identifies whether the available gene sequence is partial or complete. The "Type of Sequence" allows users to choose whether the sequence arises from cDNA (for ORL11), from mRNA or if it is a pseudogene.
9. Links to the odorant receptors that are known to excite the receptors are listed for the attribute "Ligand." These are linked to OdorDB (see Subheading 3.2), a companion resource to ORDB. Likewise, the value for Molecular Models, will link to information related to computational modeling studies for that receptor in ORModelDB. Both OdorDB and ORModelDB are discussed in detail in the next two sections.

3.2 OdorDB

1. In ORDB, (Fig. 1), odors are listed as "values" for the "ligand" attribute: heptanal and octanal. These two odorants have been functionally characterized as eliciting an excitatory response in the rat I7 receptor. When the link for "heptanal" is clicked, it takes the user to a page dedicated to this odorant (Fig. 2).
2. The attributes for octanal are as follows: Name, Chemical Abstracts Service (CAS) number, its molecular formula, whether it has a cyclic structure, its hydrocarbon feature (straight chain, branched chain, saturated, unsaturated, etc.), and the two-dimensional structure. If a Glomerular activity map, that illustrated response to that odor, was available in OdorMapDB, another database within SenseLab (http://senselab.med.yale.edu/OdorMapDB/), a relevant link is provided.
3. OdorDB (senselab.med.yale.edu/odordb) is a repository of odorants. Currently, 250 odorants are stored in this resource. They bear the same attributes as those for octanal, mentioned in the paragraph above. Figure 3 shows the same page as seen

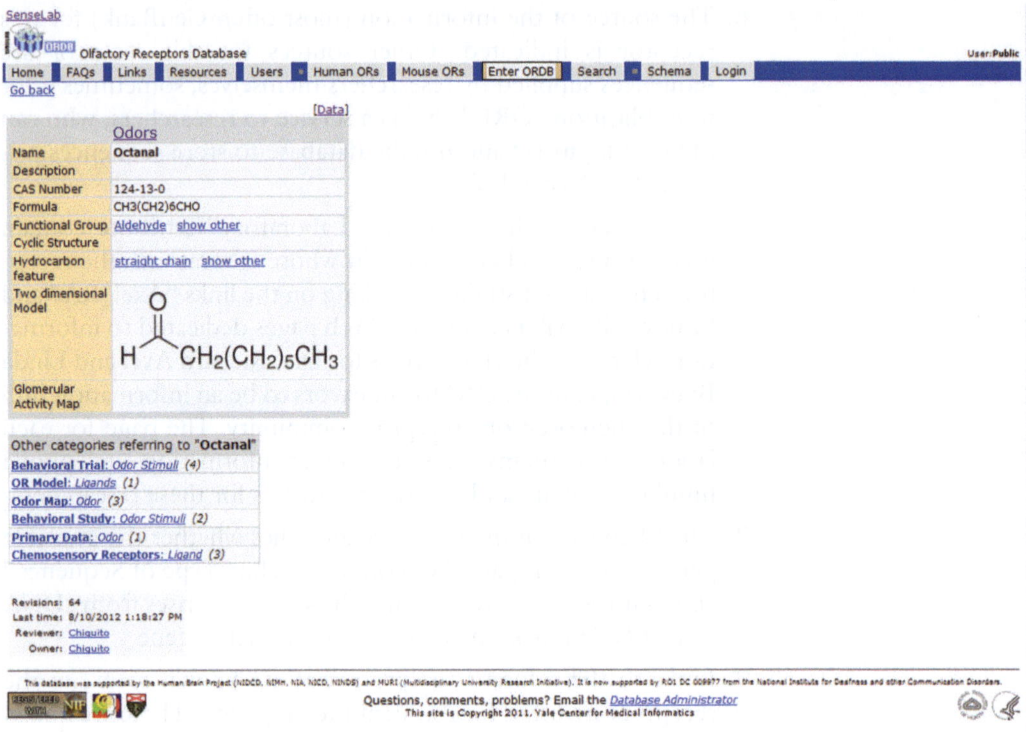

Fig. 2 The Web page for octanal in ORDB which results from clicking the "octanal" link. This is a value for the "ligand" attribute, as seen in Fig. 1. The attributes are on the *right hand side* and the values are on the *left hand side*. Values stored as classes are seen as links

in Fig. 2. The different colors and specific links, however, indicate that the same information is accessible from a different resource (pointing to *data integration, discussed in several instances in this chapter*). The link seen in the list of links at the top of the page in Fig. 3, "O.R. by ligand." It identifies Odors by the ORs with which they are known to interact. Figure 4 shows a truncated portion of this page. 97 odorant molecules have been identified as binding and exciting ORs.

4. In Fig. 4, the middle column provides links to the odorant molecules. These descriptions have been shown in Figs. 2 and 3. The right hand column depicts the number of ORs that are excited by these an odorant in the same row. For example, (+)-Dihydrocarvone is an odorant that, according to ORDB, has been functionally assessed against ten ORs. By clicking the number "10" in the right hand column a list of these ORs in ORDB can be accessed.

3.3 ORModelDB

1. ORModelDB is the newest database in SenseLab. It was created to disseminate information related to the computational studies

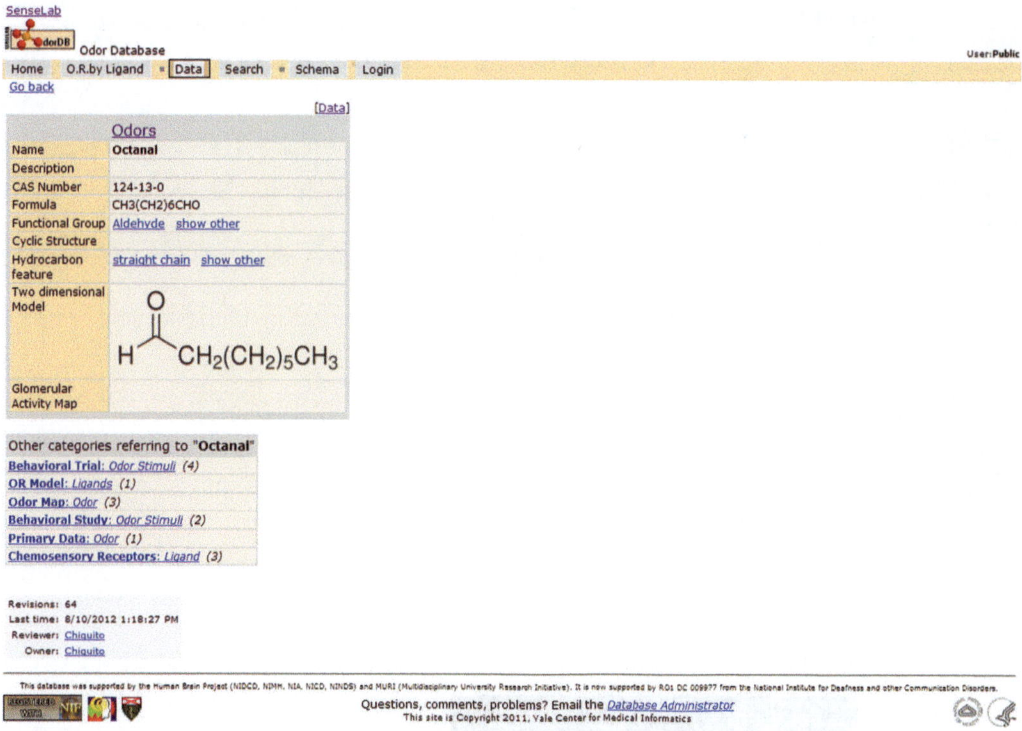

Fig. 3 The figure represents the octanal odor stored in OdorDB. This representation, other than the color differences and the links towards the *top* of the page, is the same as the ligand representation for "octanal" in ORDB, as seen in Fig. 2

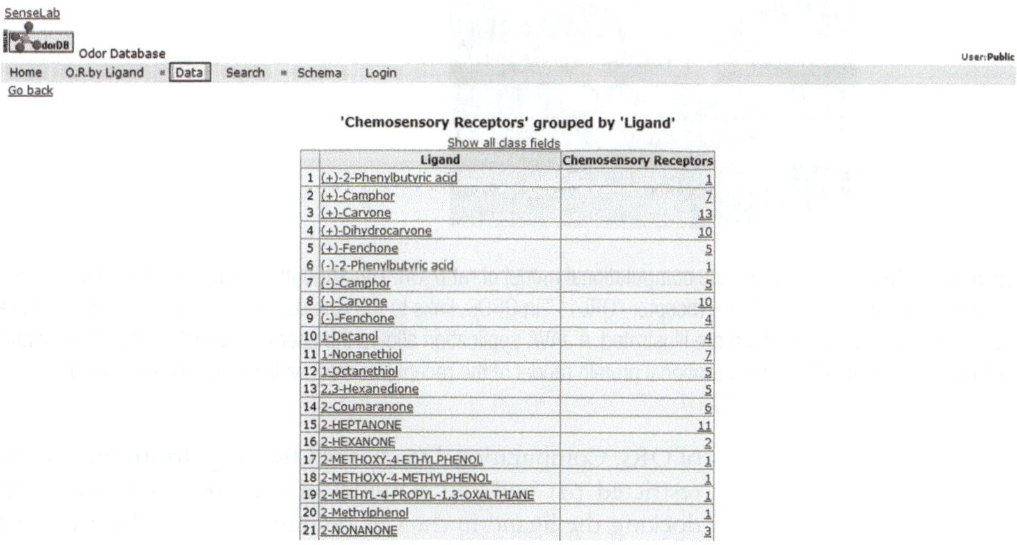

	Ligand	Chemosensory Receptors
1	(+)-2-Phenylbutyric acid	1
2	(+)-Camphor	7
3	(+)-Carvone	13
4	(+)-Dihydrocarvone	10
5	(+)-Fenchone	5
6	(-)-2-Phenylbutyric acid	1
7	(-)-Camphor	5
8	(-)-Carvone	10
9	(-)-Fenchone	4
10	1-Decanol	4
11	1-Nonanethiol	7
12	1-Octanethiol	5
13	2,3-Hexanedione	5
14	2-Coumaranone	6
15	2-HEPTANONE	11
16	2-HEXANONE	2
17	2-METHOXY-4-ETHYLPHENOL	1
18	2-METHOXY-4-METHYLPHENOL	1
19	2-METHYL-4-PROPYL-1,3-OXALTHIANE	1
20	2-Methylphenol	1
21	2-NONANONE	3

Fig. 4 The figure represents a list of 97 (out of 250) odors stored in OdorDB. The (truncated) list illustrated in the figure is one of odors that are known to excite or otherwise interact with olfactory receptors. By clicking on the odor name, the user is taken to information of the odor as seen in Figs. 2 and 3. By clicking the number on the *right hand side* of the table, which represents the number of ORs that are excited by each odor, the user can access these ORs in ORDB

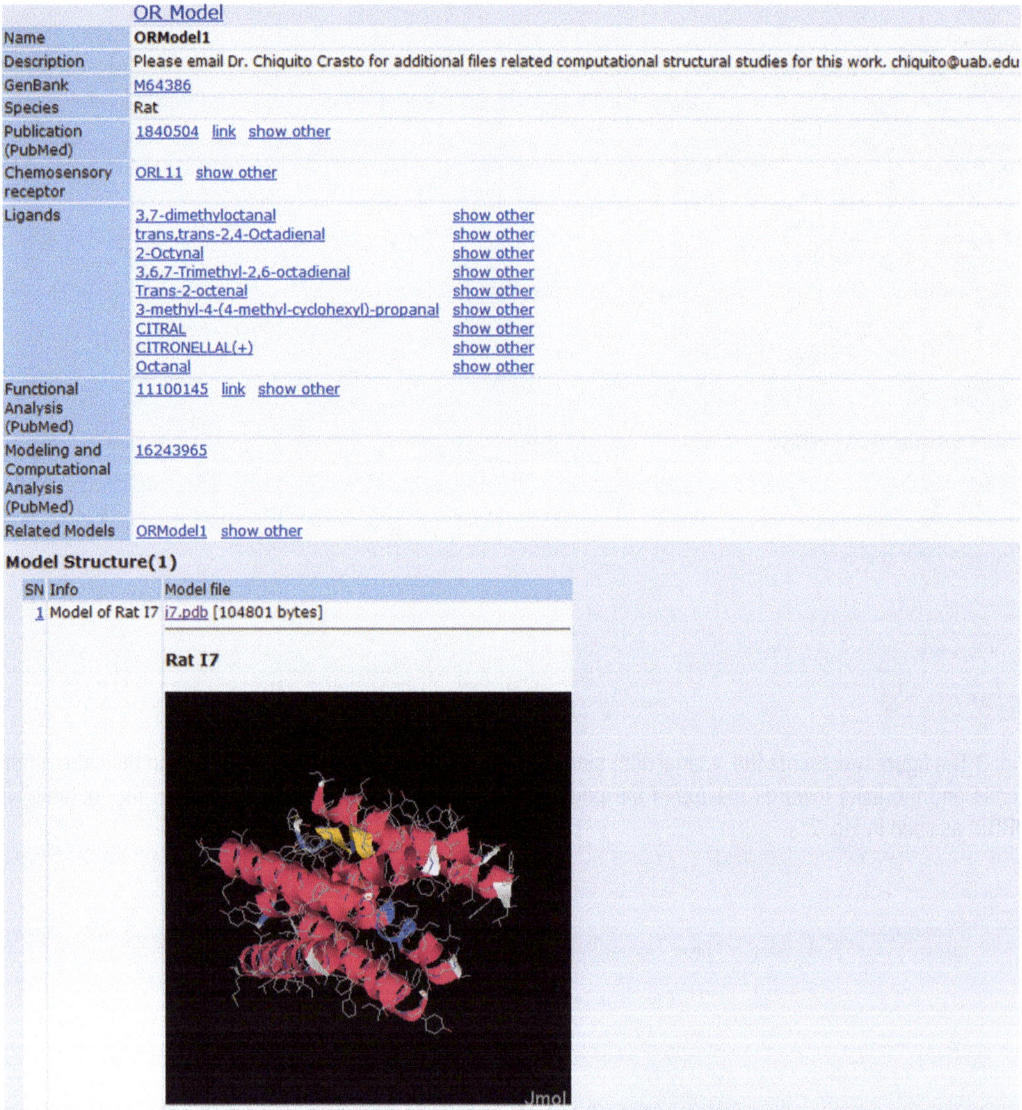

Fig. 5 The ORModel DB entry for the computational model of rat I7 OR. The attributes for this entry are listed in *blue* and the values in *gray*. Links to the receptor "ORL11" in ORDB, links to the ligands in OdorDB, links to the GenBank source of the gene, and PubMed are illustrated. A JAVA application allows the user to manipulate the model within the browser. The pdb of the computational protein model of the receptor is also available for free download

of ORs. Computational studies of OR, range from (but are not restricted to) building a model of the OR, computationally docking the ligand in the OR binding pocket and performing dynamic simulation studies to study the interactions of the odorant and the OR, all with a view to providing a mechanistic basis for OR function.

2. The attributes or descriptors of an ORModel (Fig. 5) include a link to the source of the sequence of that receptor (typically

from GenBank), links to articles that describe the identification of the OR, the results of functional analyses and those that describe the computational studies involving the OR. The link containing the ORDB name allows the user to access additional information about the OR in ORDB. Odor ligands associated with this OR are identified as links which describe the odorants in OdorDB.

3. A new feature specific to ORModelDB is the incorporation of a JAVA-based viewer that allows users to view and manipulate the structure of the OR within the Web browser. ORModelDB was created, in part, to share structural information related to ORs. The PDB (protein data bank) formatted file of the computational model is also available for download by a user, as are any other relevant files that provide additional information related to the computational studies. The "Description" field for an ORModelDB entry contains an e-mail address of the principal investigator of the modeling work so that interested users can contact him or her—for additional information about the modeling process, sharing of files (not included as a link for download), etc.

3.4 Database Architecture

1. All of SenseLab's databases are based on the EAV/CR (Entity-Attribute-Value with Classes and Relationships) database architecture (60–62). Figures 1, 2 and 3 are illustrative of this. EAV/CR was designed as a "metadata-driven infrastructure:" (63) it is possible to integrate heterogeneous databases within its framework. Representing resources using the EAV/CR framework precludes the necessity of re-storing the same information in different databases.
2. In Fig. 1, for example, within the EAV/CR framework, the ORDB database consists of several classes, and the entries for ORs are entities within the Chemosensory Receptor class. The mode of representation of a class is exactly like that of an entity. Each OR entry—an Entity-Class—is also called an Object. The properties that define an entity are attributes of a class. Attributes are qualified by specific values.
3. In ORDB, as seen in Fig. 1, the attributes are in the left hand side column of the figure and the corresponding values are in the right hand side column. The real power behind the EAV/CR framework can be seen in some of the "Values" for an OR. These values are highlighted as links. For example, in Fig. 1, if a user clicks "Buck_L" which is a value for Sequence Lab—denoting that Dr. Linda Buck's research was responsible for the early developments related to ORL11 (rat OR I7)—the user is taken to a page which depicts Buck_L as an independent entity defined by a separate class. And as an independent entity, the (researcher's) Name has its own attributes and values: "Affiliation," "Mailing

Address," "Telephone Number," etc. The information for each of these attributes is then a value.

The curators of ORDB therefore, have an easier task when entering an OR into the database. When entering a "Value" (researchers' names for the "Sequence Laboratory" attribute, curators can select from a list of researcher names. For every class, a link "show other" is seen. For the "Buck_L" Sequence Laboratory example, clicking on "Show Other" will result in a list of all other ORDB entries which have the "Buck_L" Sequence Laboratory assignation.

4. The same is true for all other values that are stored as classes. From an algorithm development standpoint, the same script is used to process every class in a database. Note the close resemblance, especially in color between Figs. 1 and 2. The left hand column is orange; the right hand column is gray. When a class is invoked by a user clicking the link for that class, the same program-script, which is created to present information about that class, accesses the databases, accesses all the attributes associated with the class in the attributes table and extracts all the corresponding values. The attribute-value dyads are then presented to the user in a Web page in tabulated format. Thus for the OR class—the results as seen in Fig. 1 are extracted and presented, likewise for "heptanal" in Fig. 2.

3.5 Database Integration and Interoperability

1. EAV/CR provides mechanisms that facilitate data sharing between all of SenseLab's databases, and indeed, ORDB, OdorDB, and ORModelDB. These mechanisms are defined by restriction levels. Elements in some classes may be shared as links that connect one database to another, information can be entered for one resource by a curator of another resource, and even schemas can altered by curators of other databases inside an EAV/CR data storage system. The flexibility afforded by the database architecture, whose applicability in the ORDBs was illustrated in Subheading 3.4.
2. The basis for data-integration in all of SenseLab's databases lies in the universally applied database schema. All the database information arises from the same storage system. This includes the metadata, schema, and semantic annotations. All EAV/CR data integration routines are metadata-driven.
3. Figure 6 is the XML representation for ORL11 in Fig. 1. By clicking the "Data" link for every class in SenseLab databases, the XML schema for that entry is revealed. In the figure, "c" preceding a number denotes that it is a class. In this case, the class is OR (Olfactory or Chemosensory Receptor); "o" before a number denotes an object or a class; "a" before a number denotes an attribute. "o559" is associated with the entity ORL11—the specific entry for the rat I7 OR (Fig. 1). There is

```
▼<data time="0.0312473999947542">
 ▼<data_database db_id="d4">
  ▼<data_class class_id="c22" version="5">
   ▼<object id="o559" name="ORL11" description="" umls_cui="" version="22" version_date="8/27/2012 11:15:15 PM">
     <att_value att_id="a80" value="o1364" value_object_name="ORL (Olfactory Receptor Like)"/>
     <att_value att_id="a30" value="o804" serial_no="0" value_object_name="Rattus norvegicus (rat)"/>
     <att_value att_id="a31" value="o821" value_object_name="olfactory epithelium"/>
     <att_value att_id="a32" value="11"/>
     <att_value att_id="a45" value="o131893" value_object_name="M64386"/>
     <att_value att_id="a44" value="o129268" serial_no="0" value_object_name="1840504"/>
     <att_value att_id="a35" value="I7"/>
     <att_value att_id="a36" value="o835" serial_no="0" value_object_name="GENBANK"/>
     <att_value att_id="a37" value="o848" serial_no="0" value_object_name="AXEL_R"/>
     <att_value att_id="a37" value="o855" serial_no="0" value_object_name="BUCK_L"/>
     <att_value att_id="a39" value="o1362" value_object_name="Full-Length"/>
     <att_value att_id="a82" value="o144593" serial_no="1" value_object_name="Heptanal"/>
     <att_value att_id="a82" value="o124" serial_no="2" value_object_name="Octanal"/>
     <att_value att_id="a40"
     value="ATGGAGCGAAGGAACCACAGTGGGAGAGTGAGTGAATTTGTGTTGCTGGGTTTCCCAGCTCCTGCCCCACTGCGAGTACTACTATTTTTCCTTTCTCTTCTGGNCTATGTGTTGGT
     <att_value att_id="a41"
     value="MERRNHSGRVSEFVLLGFPAPAPLRVLLFFLSLLXYVLVLTENMLIIIAIRNHPTLHKPMYFFLANMSFLEIWYVTVTIPKMLAGFIGSKENHGQLISFEACMTQLYFFLGLGCTE
     <att_value att_id="a42" value="o842" value_object_name="cDNA"/>
     <att_value att_id="a99" value="o143600" value_object_name="ORModel1"/>
     <att_value att_id="a214" value="10781022"/>
     <att_value att_id="a338" value="559"/>
   </object>
  </data_class>
 </data_database>
</data>
```


Fig. 6 The figure illustrates the XML representation for ORL11 in ORDB (Fig. 1). A XML file is available for every entry in the olfactory databases in SenseLab. The file shows the object identifiers of the entries when stored as classes, the attribute identifiers of attributes, and the formats for each of the values, e.g., integer for chromosome number, text for common name, and large text or memo for nucleotide and amino acid sequences

no other class in the entire SenseLab schema with the assignation "o559." The attribute "a80" is for the attribute "CR Type"; this denotes whether the CR is an OR, taste receptor, pheromone receptor, etc. ORL, because it has multiple instances, is stored as an object (much like octanal or Buck_L in Subheadings 3.1 and 3.2) "o1364." Only when a value is stored as a class does it receive an object identifier.

4. "a32" denotes the attribute "Chromosome." Since this is a unique number, it is stored as a textual value (as an Integer) and not given a class designation. This is the same for attributes "a40" and "a41," for "Nucelotides" and "Amino Acid" sequences, which are unique instances of text.
5. If these data are to be exchanged or represented differently (through OdorDB instead of ORDB), the script has to be merely modified, but is still designed to access the same tables and instances in the database.

   Consider the following three URLs: http://senselab.med.yale.edu/ORDB/eavData.aspx?o=144593, http://senselab.med.yale.edu/odordb/eavData.aspx?o=144593 and http://senselab.med.yale.edu/ORModelDB/eavData.aspx?o=144593. Each represents pages in ORDB, OdorDb and ORModelDB, respectively (Figs. 2, 3 and 5). Class (or entity) with unique identifier "o144593" represents in the database, the odorant "Heptanal." All three of the above URLs invoke the eavData.aspx (Active Server Pages C#) script that extracts information from the SenseLab databases and presents it to the user in a dynamically created Web page. Clicking on the first two links results in Web pages as seen in Figs. 2 and 3, which are identical in every respect, except for the color schemes of the pages. One represents ORDB; the other represents OdorDB. The third URL represents the exact same information in the same format as seen in Figs. 2 and 3, but with the color schemes of ORModelDB, seen in Fig. 5.

6. If the "o=144593" in the above URLs were replaced by "o=559," the rest of the URL remaining the same, Fig. 1 would result. The eavData.aspx script is universal for all of SenseLab. When this script is deployed, for a specific class (represented by the object identifier), it scans the database identifying the attributes and values and presents it in the result Web page.
7. The question as to why different databases are needed can be answered thus: A user who seeks information about Olfactory (or other chemosensory) receptors can extract the information about ORs and click on links of choice. If, on the other hand, a user who seeks information about odor molecules, with no interest in receptors, can visit OdorDB, and access information about a specific odor that is stored in SenseLab; likewise, a user interested in computational models of ORs and not in other aspects of ORs, can find that information in ORModelDB. This integration-leveraging database infrastructure design ensures that data does not have to be replicated and stored in different databases.

3.6 Multiple-Parameter Search

1. One of the challenges of databases that house large amounts of data in complex database structures and hierarchies is to be able to retrieve information that is specific and relevant to the user. Indeed, scanning through almost 15,000 entries to identify a few receptors would be a daunting task, involving scrolling through long lists and clicking on each entry before identifying the ones that were relevant.
2. Figure 7 represents a search system that, once again, leverages the database architecture of SenseLab. It is possible to search for an entry associated with any class within the database and search in such a way as to systematically filter through every attribute (to that class) to narrow search results. In Fig. 7 is represented a search for the class "chemosensory receptors." This means that the user is searching among entries in ORDB. The user chooses "CR Type" as ORL.
3. The yellow tab marked "Look Up" indicates that an attribute-value is stored as a class. The tab is designed to aid the user avoid having to type the specific value. For the CR Type attribute, if the user clicks on this tab, a pop-up window with all the possible choices for this attribute, ORL, TPR, VNR, etc., will be available. The user can click on the choice (or more

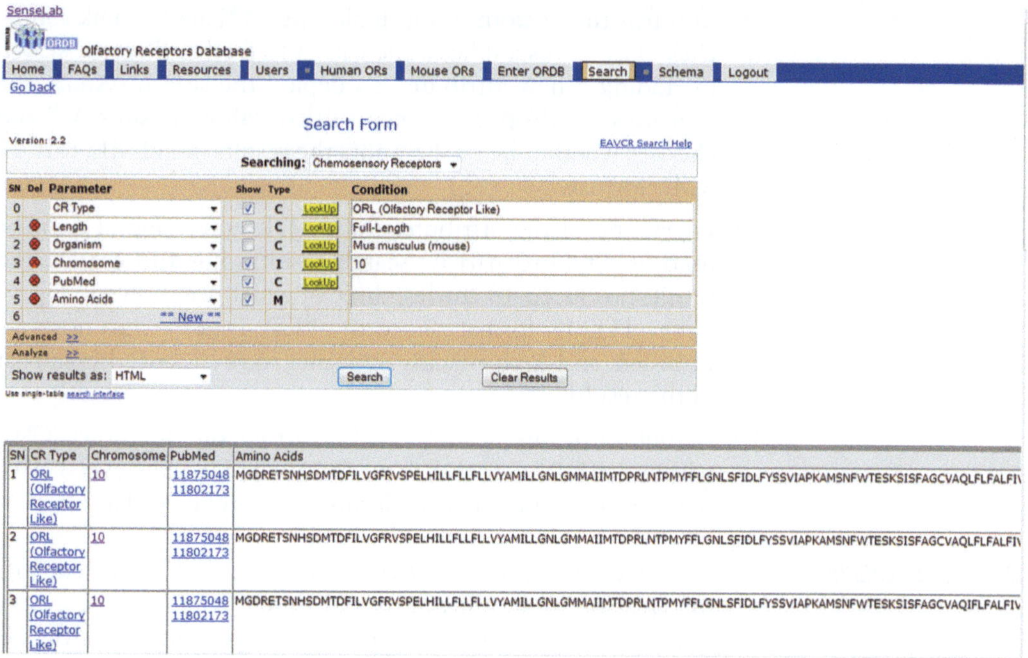

Fig. 7 Shows the multiple parameter search system for ORDB (and available for OdorDB and ORModelDB). The form in the top part of the figure allows searching for an entity for every attribute associated with the entity. Each attribute can be used as a filter to narrow and focus the search results. The **New** link allows the introduction of a new Search filter. The *red circle* with an *x* allows the user to remove a search filter. The *check box* under "show" allows the user to opt to display that attribute in the search results. If the value that can be used as filter is stored as a class, a *look-up* table allows the user to click the specific value. The *lower part* of the figure represents the results. The query asked for a list of olfactory receptors on chromosome 10 of mouse, and asked that the PubMed links for each receptor and the amino acid sequences be reported in the HTML format

than one choice) of value and it gets entered into the field under the heading condition.

4. The user has, in the example illustrated in the figure, also chosen only full-length sequences, and organism "mouse," and receptors that are on chromosome 10. The user has chosen PubMed, but not as a filter, because no PubMed Identifier was entered in the field. The user requests that the PubMed IDs for all the relevant receptors appear in the results.
5. The user also wants all the protein sequences to appear in the results.
6. In the Column under the heading "Show" the check-box asks whether that specific information needs to be revealed in the results. For *Length* and *Organism*, the user would prefer to use these values as filters, but not have them cluttering the results—the organism and the length being already known.
7. The red circle with the "x" mark through it on the under the column heading "Del" when clicked will remove that attribute from the search filter.

8. Towards the bottom of this table, the "**New**" link can be clicked by the user if he or she wishes to further filter the search by adding a new attribute. To deploy the search system, the user uses the drop-down menu titled "Show Results As" and chooses whether he or she wants the results in HTML, text, or XML format.
9. Every one of the attributes for an OR entry (Fig. 1) can be used as filters to narrow search results. For this search, the results produced 48 entries, three of which are shown in Fig. 7, in the HTML format. Since the "CR Type," Chromosome, PubMed, and Amino Acids attributes were clicked, they appear in the results.
10. A multi-attribute search can be carried out for every Class within ORDB, OdorDB, and ORModelDB. For every class, the attributes associated with this class can be used as filters.

3.7 Auto-Population

1. The size and scope of ORDB and the need to disseminate information in an efficacious and timely manner to serve the chemosensory community and knowledge seekers worldwide, necessitates the population of ORDB using automated methods.
2. AutoPop (11) was developed as a Web-based ORDB automatic population tool. AutoPop leveraged the notion of interoperability. Interoperability is a term that has become increasingly commonplace in the domain of online information dissemination (64–66). Interoperability involves the exchange of information between two independent, disparate and heterogeneous resources. AutoPop parsed the html files of sources of OR information and extracted only the information specific to ORDB. The AutoPop front page was a Web page with a single text-field. The curator entered a GenBank Accession ID into this field. This ID was embedded in a URL that represented the GenBank page for this ID. A PERL (Practical Extraction and Report Language) was written to access the GenBank servers and download the Web page to the SenseLab servers for additional processing.
3. An ORDB knowledgebase was created, which consisted of a representation of the schema of the database, containing lists of unique attribute and object Ids of the values of attributes stored in ORDB as independent classes. The value descriptors of these attributes and objects were also stored in this knowledgebase.
4. During the processing of the HTML page (also known as screen scraping—a subsequently discontinued process) if the receptor gene sequence came from an organism, e.g., *Mus musculus*, and if the organism was identified as such, it was matched against the ORDB knowledgebase and mapped onto

object-value "o144" (for mouse) and attribute "a30" (organism). Since the XML file was dynamically generated while the HTML page downloaded from GenBank was being processed, a line <a30 object_name='mus musculus'>o144<a30/> was written to the XML file. For an entry for a mouse OR gene, if the strain, e.g., 129/SVJ was identified, it was mapped to the attribute for Strain "a106." Since strain is stored as a value and not an object, the line <a106 value='Strain'>129/SVJ<a106/> was added to the XML file. This file was created as each relevant attribute was identified from the source of OR information.

5. This dynamically written XML file is obtained by clicking the "Data" link on the Web page for every class in the olfactory databases. This XML file was then embedded in a Web page. Clicking "Submit" in that Web page deposited this information into ORDB directly.
6. Prior to the creation of AutoPop, for each entry, the curator cut and paste relevant information into a form, which was submitted into ORDB. Manually creating an ORDB entry took between 5 and 10 min. With AutoPop, the time from entering the GenBank Accession ID to the creation of the html form file with XML file embedded was less than 10 s. This intermediate step was deemed necessary so that the curator could validate information being added was accurate before submitting it to ORDB.
7. AutoPop was not necessarily restricted to populating ORDB one receptor at a time. As long as an XML file was created, the nested hierarchies allowed multiple entries. This was tested for the EXPASY (http://expasy.org/) (formerly, Swiss-Prot) resource. A script was written that processed EXPASY protein entries. 74 OR proteins were identified and information for all the ORs were written to a single XML file. All 74 protein sequences were deposited into ORDB in less than 30 s.
8. Creating AutoPop as a Web-based system ensured that the curator could populate the database remotely.

3.8 Updating ORDB

1. The screen scraping methodology is less than ideal. The effort required in writing a program that parses an HTML file—which is designed for ease of human viewing—to extract information that should be processed by a computer poses many difficulties.
2. Researchers at NLM, the purveyors of information dissemination at NCBI, made gene information available through the XML format for every GenBank entry. The XML formatted representation of a GenBank entry has now been discontinued. The ideal format available for machine-readable extraction is the ASN.1 (Abstract Syntax Notation. One) format.

3. One challenge is to populate and update ORDB when new information becomes available (Subheading 3.5). Another is to modify and update ORDB entries as the source information evolves. ORDB curators were confronted with the knowledge that as more publications related to olfactory genes were identified, GenBank entries for these genes were updated to reflect multiple links to PubMed.
4. When ORDB was first being populated, one PubMed link existed per gene. To update ORDB and ensure that all PubMed links that are in a GenBank entry for an OR gene are included, ORDB curators developed algorithms that specifically targeted all the sources of olfactory genes in GenBank to identify new PubMed links for the genes and update ORDB to reflect this.
5. Researchers at the National Library of Medicine have created e-utils (67), a system that allows users to search, access, and download large amounts of data from all NCBI sources. There are several resources in e-utils. Here we used two: E-Search and E-Fetch. E-utils framework involves the design of specific URLs to access NCBI resources in a customized manner. These URLs are not very different from those a user would enter into a Web browser window. For efficient use of the e-utils framework, we developed PERL scripts.
6. We first used the ORDB search feature described in the previous section and illustrated in Fig. 7 to perform a search on all ORDB, retrieving for every gene, the ORDB object identifiers, the ORDB name, and the GenBank Accession ID.

   E-search: Our e-search URL was http://eutils.ncbi.nlm.nih.gov/entrez/eutils/esearch.fcgi?db=nucleotide&term=NW_000328&usehistory=y. The URL shows that we access the e-utils resources, the nucleotide database, and the GenBank Accession ID "NW_00328." On running the e-search program, an XML file is generated which contains the unique identifier for this gene in GenBank. We created an XML parser that processed this XML file to extract this unique identifier.

   E-Fetch: The unique identifier (for illustrative purposes UID) extracted from the E-search result was then embedded in the following URL: http://www.ncbi.nlm.nih.gov/entrez/eutils/efetch.fcgi?db=nucleotide&id=UID&rettype=ASN.1&retmode=text. This URL indicates that we use the e-fetch feature in e-utils, access the "nucleotide" database, and retrieve the file for this gene in the ASN.1 format as a text file. Another script was written to process this results file and extract only the PubMed Identifiers.
7. The above three steps were cycled for all 14,500 entries in ORDB. The process, running on a personal work station was completed in 40 h. Each of the PubMed identifiers were

mapped to the object IDs and then this information was deposited in ORDB. 7000 entries, close to 50% of the entries had more than one PubMed link, reflecting research completed and published that was related to each OR gene—after the ORDB entry had been made (with one PubMed link).

Since the above steps are semi-automated, deploying these every few months will ensure that ORDB is continually updated.

4 Notes

1. Editing, curating, and algorithmic development that facilitates the above functionality of SenseLab's olfactory databases is only within the purview of those with special access and administrative privileges.
2. ORDB, OdorDB, and ORModelDB allow users to enter, edit, and update entries with limited user privileges.
3. Users can use the available database infrastructure to store and access information privately, until an associated article is published or the information is ready to be made publicly available.

Acknowledgments

This work is generously supported in part by grant R21DC011068, R01DC009977, and R01DA021253 from the National Institutes of Health.

References

1. Buck L, Axel R (1991) A novel multigene family may encode odorant receptors: a molecular basis for odor recognition. Cell 65:175–187
2. Beets MG (1970) The molecular parameters of olfactory response. Pharmacol Rev 22:1–34
3. Mateson JF (1954) The olfactory area and the olfactory receptor process. Ann N Y Acad Sci 58:83–95
4. Mombaerts P (2004) Love at first smell–the 2004 Nobel Prize in physiology or medicine. N Engl J Med 351:2579–2580
5. Zhang X, Firestein S (2002) The olfactory receptor gene superfamily of the mouse. Nat Neurosci 5:124–133
6. Sharon D, Glusman G, Pilpel Y, Horn-Saban S, Lancet D (1998) Genome dynamics, evolution, and protein modeling in the olfactory receptor gene superfamily. Ann N Y Acad Sci 855: 182–193
7. Glusman G, Bahar A, Sharon D, Pilpel Y, White J, Lancet D (2000) The olfactory receptor gene superfamily: data mining, classification, and nomenclature. Mamm Genome 11:1016–1023
8. Glusman G, Yanai I, Rubin I, Lancet D (2001) The complete human olfactory subgenome. Genome Res 11:685–702
9. Gilad Y, Man O, Glusman G (2005) A comparison of the human and chimpanzee olfactory receptor gene repertoires. Genome Res 15:224–230
10. Crasto C, Singer MS, Shepherd GM (2001) The olfactory receptor family album. Genome Biol 2:REVIEWS1027

11. Crasto C, Marenco L, Miller P, Shepherd G (2002) Olfactory receptor database: a metadata-driven automated population from sources of gene and protein sequences. Nucleic Acids Res 30:354–360
12. Crasto CJ, Marenco LN, Liu N, Morse TM, Cheung KH, Lai PC, Bahl G, Masiar P, Lam HY, Lim E, Chen H, Nadkarni P, Migliore M, Miller PL, Shepherd GM (2007) SenseLab: new developments in disseminating neuroscience information. Brief Bioinform 8:150–162
13. Healy MD, Smith JE, Singer MS, Nadkarni PM, Skoufos E, Miller PL, Shepherd GM (1997) Olfactory receptor database (ORDB): a resource for sharing and analyzing published and unpublished data. Chem Senses 22:321–326
14. Shepherd GM, Mirsky JS, Healy MD, Singer MS, Skoufos E, Hines MS, Nadkarni PM, Miller PL (1998) The human brain project: neuroinformatics tools for integrating, searching and modeling multidisciplinary neuroscience data. Trends Neurosci 21:460–468
15. Skoufos E, Healy MD, Singer MS, Nadkarni PM, Miller PL, Shepherd GM (1999) Olfactory receptor database: a database of the largest eukaryotic gene family. Nucleic Acids Res 27:343–345
16. Skoufos E, Marenco L, Nadkarni PM, Miller PL, Shepherd GM (2000) Olfactory receptor database: a sensory chemoreceptor resource. Nucleic Acids Res 28:341–343
17. Osterhammel P, Terkildsen K, Zilstorff K (1969) Electro-olfactograms in man. J Laryngol Otol 83:731–733
18. Hummel T, Knecht M, Kobal G (1996) Peripherally obtained electrophysiological responses to olfactory stimulation in man: electro-olfactograms exhibit a smaller degree of desensitization compared with subjective intensity estimates. Brain Res 717:160–164
19. Scott JW, Scott-Johnson PE (2002) The electroolfactogram: a review of its history and uses. Microsc Res Tech 58:152–160
20. Knecht M, Hummel T (2004) Recording of the human electro-olfactogram. Physiol Behav 83:13–19
21. Shirokova E, Schmiedeberg K, Bedner P, Niessen H, Willecke K, Raguse JD, Meyerhof W, Krautwurst D (2005) Identification of specific ligands for orphan olfactory receptors. G protein-dependent agonism and antagonism of odorants. J Biol Chem 280:11807–11815
22. Matarazzo V, Clot-Faybesse O, Marcet B, Guiraudie-Capraz G, Atanasova B, Devauchelle G, Cerutti M, Etievant P, Ronin C (2005) Functional characterization of two human olfactory receptors expressed in the baculovirus Sf9 insect cell system. Chem Senses 30:195–207
23. Doengi M, Hirnet D, Coulon P, Pape HC, Deitmer JW, Lohr C (2009) GABA uptake-dependent Ca(2+) signaling in developing olfactory bulb astrocytes. Proc Natl Acad Sci USA 106:17570–17575
24. Murmu MS, Stinnakre J, Real E, Martin JR (2011) Calcium-stores mediate adaptation in axon terminals of olfactory receptor neurons in Drosophila. BMC Neurosci 12:105
25. Tsuboi A, Imai T, Kato HK, Matsumoto H, Igarashi KM, Suzuki M, Mori K, Sakano H (2011) Two highly homologous mouse odorant receptors encoded by tandemly-linked MOR29A and MOR29B genes respond differently to phenyl ethers. Eur J Neurosci 33:205–213
26. Breer H (1993) Olfactory receptor cells: recognition and transduction of chemical signals. Cytotechnology 11:13–16
27. Singer MS, Shepherd GM (1994) Molecular modeling of ligand-receptor interactions in the OR5 olfactory receptor. Neuroreport 5:1297–1300
28. Singer MS (2000) Analysis of the molecular basis for octanal interactions in the expressed rat 17 olfactory receptor. Chem Senses 25:155–165
29. Floriano WB, Vaidehi N, Goddard WA 3rd, Singer MS, Shepherd GM (2000) Molecular mechanisms underlying differential odor responses of a mouse olfactory receptor. Proc Natl Acad Sci USA 97:10712–10716
30. Zhao H, Ivic L, Otaki JM, Hashimoto M, Mikoshiba K, Firestein S (1998) Functional expression of a mammalian odorant receptor. Science 279:237–242
31. Araneda RC, Kini AD, Firestein S (2000) The molecular receptive range of an odorant receptor. Nat Neurosci 3:1248–1255
32. Bozza T, Feinstein P, Zheng C, Mombaerts P (2002) Odorant receptor expression defines functional units in the mouse olfactory system. J Neurosci 22:3033–3043
33. Luu P, Acher F, Bertrand HO, Fan J, Ngai J (2004) Molecular determinants of ligand selectivity in a vertebrate odorant receptor. J Neurosci 24:10128–10137
34. Chase J, Touhara K, Prestwich GD, Schal C, Blomquist GJ (1992) Biosynthesis and endocrine control of the production of the German cockroach sex pheromone 3,11-dimethylnonacosan-2-one. Proc Natl Acad Sci USA 89:6050–6054

35. Touhara K, Sengoku S, Inaki K, Tsuboi A, Hirono J, Sato T, Sakano H, Haga T (1999) Functional identification and reconstitution of an odorant receptor in single olfactory neurons. Proc Natl Acad Sci USA 96:4040–4045
36. Touhara K (2001) Functional cloning and reconstitution of vertebrate odorant receptors. Life Sci 68:2199–2206
37. Touhara K (2002) Odor discrimination by G protein-coupled olfactory receptors. Microsc Res Tech 58:135–141
38. Katada S, Nakagawa T, Kataoka H, Touhara K (2003) Odorant response assays for a heterologously expressed olfactory receptor. Biochem Biophys Res Commun 305:964–969
39. Oka Y, Nakamura A, Watanabe H, Touhara K (2004) An odorant derivative as an antagonist for an olfactory receptor. Chem Senses 29:815–822
40. Oka Y, Omura M, Kataoka H, Touhara K (2004) Olfactory receptor antagonism between odorants. EMBO J 23:120–126
41. Oka Y, Katada S, Omura M, Suwa M, Yoshihara Y, Touhara K (2006) Odorant receptor map in the mouse olfactory bulb: in vivo sensitivity and specificity of receptor-defined glomeruli. Neuron 52:857–869
42. Touhara K (2007) Deorphanizing vertebrate olfactory receptors: recent advances in odorant-response assays. Neurochem Int 51:132–139
43. Schmiedeberg K, Shirokova E, Weber HP, Schilling B, Meyerhof W, Krautwurst D (2007) Structural determinants of odorant recognition by the human olfactory receptors OR1A1 and OR1A2. J Struct Biol 159:400–412
44. Grosmaitre X, Fuss SH, Lee AC, Adipietro KA, Matsunami H, Mombaerts P, Ma M (2009) SR1, a mouse odorant receptor with an unusually broad response profile. J Neurosci 29:14545–14552
45. Nara K, Saraiva LR, Ye X, Buck LB (2011) A large-scale analysis of odor coding in the olfactory epithelium. J Neurosci 31:9179–9191
46. Muller G (2000) Towards 3D structures of G protein-coupled receptors: a multidisciplinary approach. Curr Med Chem 7:861–888
47. Lai PC, Bahl G, Gremigni M, Matarazzo V, Clot-Faybesse O, Ronin C, Crasto CJ (2008) An olfactory receptor pseudogene whose function emerged in humans: a case study in the evolution of structure-function in GPCRs. J Struct Funct Genomics 9:29–40
48. Lai PC, Singer MS, Crasto CJ (2005) Structural activation pathways from dynamic olfactory receptor-odorant interactions. Chem Senses 30:781–792
49. Lai PC, Crasto CJ (2012) Beyond modeling: all-atom olfactory receptor model simulations. Front Genet 3:61
50. Vaidehi N, Floriano WB, Trabanino R, Hall SE, Freddolino P, Choi EJ, Zamanakos G, Goddard WA 3rd (2002) Prediction of structure and function of G protein-coupled receptors. Proc Natl Acad Sci USA 99:12622–12627
51. Hall SE, Floriano WB, Vaidehi N, Goddard WA 3rd (2004) Predicted 3-D structures for mouse I7 and rat I7 olfactory receptors and comparison of predicted odor recognition profiles with experiment. Chem Senses 29:595–616
52. Floriano WB, Vaidehi N, Goddard WA 3rd (2004) Making sense of olfaction through predictions of the 3-D structure and function of olfactory receptors. Chem Senses 29:269–290
53. Hummel P, Vaidehi N, Floriano WB, Hall SE, Goddard WA 3rd (2005) Test of the Binding Threshold Hypothesis for olfactory receptors: explanation of the differential binding of ketones to the mouse and human orthologs of olfactory receptor 912-93. Protein Sci 14: 703–710
54. Palczewski K, Kumasaka T, Hori T, Behnke CA, Motoshima H, Fox BA, Le Trong I, Teller DC, Okada T, Stenkamp RE, Yamamoto M, Miyano M (2000) Crystal structure of rhodopsin: a G protein-coupled receptor. Science 289:739–745
55. Okada T, Sugihara M, Bondar AN, Elstner M, Entel P, Buss V (2004) The retinal conformation and its environment in rhodopsin in light of a new 2.2 A crystal structure. J Mol Biol 342:571–583
56. Hanson MA, Roth CB, Jo E, Griffith MT, Scott FL, Reinhart G, Desale H, Clemons B, Cahalan SM, Schuerer SC, Sanna MG, Han GW, Kuhn P, Rosen H, Stevens RC (2012) Crystal structure of a lipid G protein-coupled receptor. Science 335:851–855
57. Shimamura T, Shiroishi M, Weyand S, Tsujimoto H, Winter G, Katritch V, Abagyan R, Cherezov V, Liu W, Han GW, Kobayashi T, Stevens RC, Iwata S (2011) Structure of the human histamine H1 receptor complex with doxepin. Nature 475:65–70
58. Wu B, Chien EY, Mol CD, Fenalti G, Liu W, Katritch V, Abagyan R, Brooun A, Wells P, Bi FC, Hamel DJ, Kuhn P, Handel TM, Cherezov V, Stevens RC (2010) Structures of the CXCR4 chemokine GPCR with small-molecule and cyclic peptide antagonists. Science 330:1066–1071
59. Rasmussen SG, Choi HJ, Rosenbaum DM, Kobilka TS, Thian FS, Edwards PC, Burghammer M, Ratnala VR, Sanishvili R,

Fischetti RF, Schertler GF, Weis WI, Kobilka BK (2007) Crystal structure of the human beta2 adrenergic G-protein-coupled receptor. Nature 450:383–387

60. Kshirsagar NA, Takle MR, Nadkarni PM, Satoskar RS (1986) Effect of standard breakfast & lunch on sulphadimidine bioavailability. Indian J Med Res 83:618–622
61. Nadkarni PM, Marenco L, Chen R, Skoufos E, Shepherd G, Miller P (1999) Organization of heterogeneous scientific data using the EAV/CR representation. J Am Med Inform Assoc 6:478–493
62. Marenco L, Tosches N, Crasto C, Shepherd G, Miller PL, Nadkarni PM (2003) Achieving evolvable Web-database bioscience applications using the EAV/CR framework: recent advances. J Am Med Inform Assoc 10:444–453
63. Nadkarni P (2012) Metadata-driven software systems in biomedicine: designing systems that can adapt to changing knowledge (Health Informatics). Springer, London, New York
64. Litwin W, Abdellatif A (1986) Multidatabase interoperability. Computer (IEEE Comput. Soc.) 19:10–18
65. Litwin W, Mark L, Roussopoulos N (1990) interoperability of multiple autonomous databases. ACM Comput Surv 22:267–293
66. Noy NF, Fergerson RW, Musen MA (2000) The knowledge model of protégé-2000: combining interoperability and flexibility. In: Dieng R, Corby O (eds) Knowledge engineering and knowledge management methods, models, and tools. Springer, Heidelberg, pp 17–32
67. NCBI (2012) Entrez programming utilities help. NCBI, Bethesda, MD

Chapter 2

HORDE: Comprehensive Resource for Olfactory Receptor Genomics

Tsviya Olender, Noam Nativ, and Doron Lancet

Abstract

Olfactory receptors (ORs) constitute the largest gene family in the mammalian genome. The existence of these proteins underlies the nature of, and variability in, odorant perception. The Human Olfactory Receptor Data Explorer (HORDE, http://genome.weizmann.ac.il/horde/) is a free online resource, which presents a complete compendium of all OR genes and pseudogenes in the genome of human and four other vertebrates. HORDE includes three parts: (1) an automated pipeline, which mines OR gene and pseudogene sequences out of complete genomes, and generates gene symbols based on sequence similarity; (2) a card generator that obtains and displays annotative information on individual ORs retrieved from external databases and relevant studies; and (3) a search engine that allows user retrieval of OR information. For human ORs, HORDE specifically addresses the universe of interindividual variation, as obtained from several sources, including whole genome sequences made possible by next-generation sequencing. This encompasses single nucleotide polymorphisms (SNP) and copy number variation (CNV), including deleterious mutational events. HORDE also hosts a number of tools designed specifically to assist in the study of OR evolution and function. In this chapter, we describe the status of HORDE (build #43). We also discuss plans for future enhancements and a road map for HORDE to become a better community-based bioinformatics tool. We highlight HORDE's role as a major research tool in the study of an expanding cohort of OR repertoires.

Key words Olfactory receptors, Database, SNP, Copy number variation, Computational data-mining

1 Introduction

Olfaction is initiated following the binding of an odorant ligand to an olfactory receptor (OR) protein, which is expressed in the ciliary membrane of olfactory neurons (1–3). ORs, the largest gene superfamily in the mammalian genome, are composed of seven transmembrane domain proteins, members of the G-(Protein)-coupled protein receptors (GPCRs) hyperfamily. A typical mammalian genome contains 1,000–1,500 OR genes, evidence for the essential role of the olfactory system to survival of many species (4–6). HORDE, the Human Olfactory Receptor Data

Chiquito J. Crasto (ed.), *Olfactory Receptors: Methods and Protocols*, Methods in Molecular Biology, vol. 1003,
DOI 10.1007/978-1-62703-377-0_2,

Exploratorium, is a resource consisting of algorithmic and data storage processes that mine, annotate, and integrate complete repertoires of vertebrate OR genes. It is mainly designed to provide insight into the evolution, structure, and function of the complete human OR repertoire. Its main mission is to facilitate the navigation and scrutiny of such large gene repertoires. Another OR database is ORDB (7) (see Chapter 1), which is broader in scope, encompassing all chemosensory receptors (olfactory, vomeronasal, and gustatory receptors), but contains less information on individual human ORs. GPCRDB (8), which extracts and stores GPCR information only from public genome sources, has an OR gene section.

HORDE was established in 2000, following the publication of the first draft of the human genome (9–11). It presented for the first time a complete human OR repertoire, the result of computational data mining. Over time, HORDE has been significantly expanded and improved (12, 13), with a mission to present a complete and nonredundant digital compendium of human ORs. OR repertoires of several other vertebrate genomes have also been mined and annotated by HORDE and added to the database.

The OR repertoires of many species are replete with pseudogenes, the result of frame-disrupting mutations that spread and got fixed in the population (14). Such a process is rationalized by the functional redundancy of OR proteins, whereby each odorant is recognized by several ORs (15–17). The presence of pseudogenes is particularly noticeable in primates, which apparently depend less on olfaction for their survival, culminating in the inactivation of ~55 % of all ORs in humans. These pseudogenes are not usually identified by the standard whole genome annotation pipelines. Notably, OR pseudogenes are not annotated even by databases devoted to pseudogenes (18), because they are not processed pseudogenes (like those formed by RNA retroposition). The HORDE specialized computational pipeline allows it to be a unique resource that includes all gene and pseudogenes for a given genome.

Although ORs have been investigated for nearly two decades (2, 19), many relevant questions have not been fully answered. One example is the genetic basis of the widespread phenotypic diversity in human olfactory faculties. This is believed to be attributable, at least in part, to genomic variations in the OR genes (6, 20, 21). Relevant studies, therefore, require an accurate and comprehensive account of the genes and their genetic variations. HORDE also contains information on expression (obtained from ESTs or microarray data), as well as genomic location and clustering of ORs, and this information is relevant to several open questions. Among these are the control of gene expression in the olfactory system, including epithelial zone-specific expression as well as expression with locus and allele exclusion (22, 23), ectopic expression in non olfactory tissues (24–26), such as sperm (27), and biased expression of certain OR genes and pseudogenes (e.g., (28)).

In this chapter we describe in detail the information presented in HORDE and its use in studies of the OR superfamily. We give the full details on the annotation pipeline, HORDE's classification methods, and the analysis routines. We also describe database design, and the tools for data retrieval.

2 Data-Mining and Database Design

HORDE is committed to retrieving, processing, and archiving of the entire nonredundant list of all human ORs. We have developed an automatic pipeline (Fig. 1) that can be reapplied whenever a new version of the human genome assembly is released. This procedure, which is also used to decipher the OR repertoires of other species, is facilitated because the coding region of the OR gene is encoded by a single exon, and because the OR protein sequences are mutually conserved and distinguishable from non-OR GPCR proteins (10). Similar approaches, with minor modifications, have been later utilized by others (29–32).

This pipeline was applied to decipher the OR repertoire of the human, dog (33), opossum (34), platypus (15), lizard (35), and zebra finch (35), and with minor changes, to define the V1R repertoire of the platypus (15).

The result of our data-mining efforts is a database where information has been currently stored in a flat file format. A user search and retrieval interface with an interface written in PERL (Practical Extraction and Report Language). A new relational MySql database is now under development (See Note 2). The new database design (Fig. 2) will be equipped with the open source CakePHP framework (http://cakephp.org/) that uses a built-in Object-Relational Mapping (ORM), thus providing a simple API (application program interface) for data storage and retrieval.

3 Description of the HORDE Database

3.1 The OR Repertoires of Human and Other Species

Table 1 shows the OR counts within HORDE (build #43) for genes and pseudogenes in the species included. For all the species, the OR superfamily is classified by sequence similarity using BLASTP into 18 families and 454 subfamilies (not all species have all OR families and subfamilies) (13). The details of the classification method were described in (10, 12, 33, 36). Briefly, a new gene is classified into the same subfamily as that of its best hit if it shows ≥60 % protein sequence identity to such target. It is classified into a new subfamily of the same family as that of its best hit if it shows

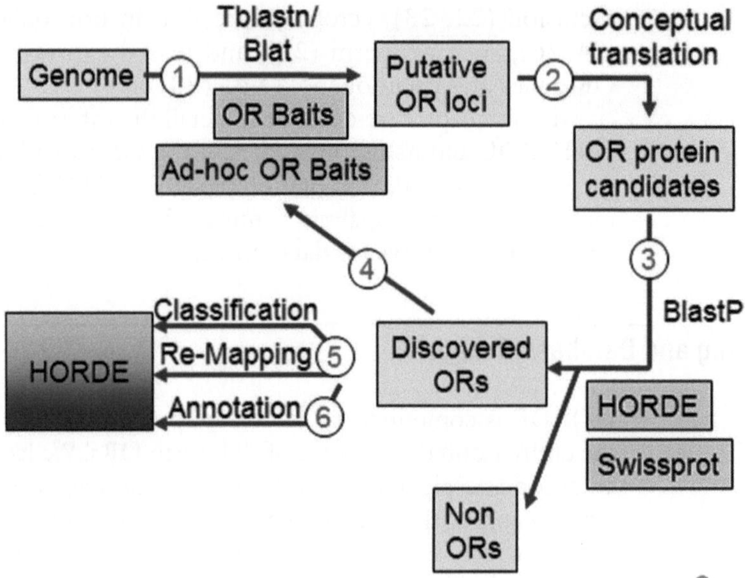

Fig. 1 The steps of the OR mining pipeline (see Note 1). (*1*) genomic screening by BLAT (62) or TBLASTN (63) with a set of "OR baits" to discover genomic putative OR loci, locations suspected to harbor OR genes. TBLASTN is preferable when mining a new genome, the faster BLAT procedure is sufficient for updating an existing repertoire. As "OR baits" we use a set of about 500 OR representatives coming from different OR subfamilies, and different vertebrates species, designed to maximally cover the OR sequence space (http://genome.weizmann.ac.il/horde/DataMining.htm). A locus is suspected as "putative OR" if the in TBLASTN $E<10^{-3}$ over at least 50 amino acids, or if it constitutes a BLAT search hit. (*2*) BioPerl is used for conceptual translation and the ORFs of 290–350 amino acids are accepted as - ORs. ORFs outside this range are suspected as OR pseudogenes and are therefore subjected to FASTY analysis to generate a "corrected" OR (64). (*3*) The resulting "OR protein candidates" are classified as ORs or non-ORs by BLASTP comparison against the current version of HORDE and against all Swissprot proteins. OR candidates which their sequence is at least 40 % identical over at least 100 amino acids to another OR are considered as OR genes (10). Borderline cases with <40 % protein identity to the best OR, with another GPCR being the best hit in EXPASY (formerly, SwissProt), are classified as non-ORs. (*4*) Steps *1–3* are repeated using newly discovered ORs as baits ("Ad-hoc OR baits"), until no novel hits are detected. (*5*) OR genes are remapped to the genome; they are classified into families and subfamilies; a unique symbol is assigned to each gene (see text). (*6*) Additional annotation is integrated into the database from external sources. The external sources currently accessed are GeneBank, dbSNP, ORDB (7), DGV (43), and BioGPS (47). Updates necessary after the publication of a new genome version for a species already included in HORDE are performed as follows: All ORs of this species undergo an update of genomic coordinates to those of the new version, using BLAT. Then the mining pipeline is reapplied; ORs already present are identified by their genomic coordinates, and newly discovered ones are added to HORDE and annotated

≥40 %, but ≤60 % protein sequence identity to such target. In cases where the similarity to the best hit is lower than ≤40 %, the gene is candidate to define a new family, but such a definition is carried out with caution, by a more elaborate procedure that involves phylogenetic and principal component analyses.

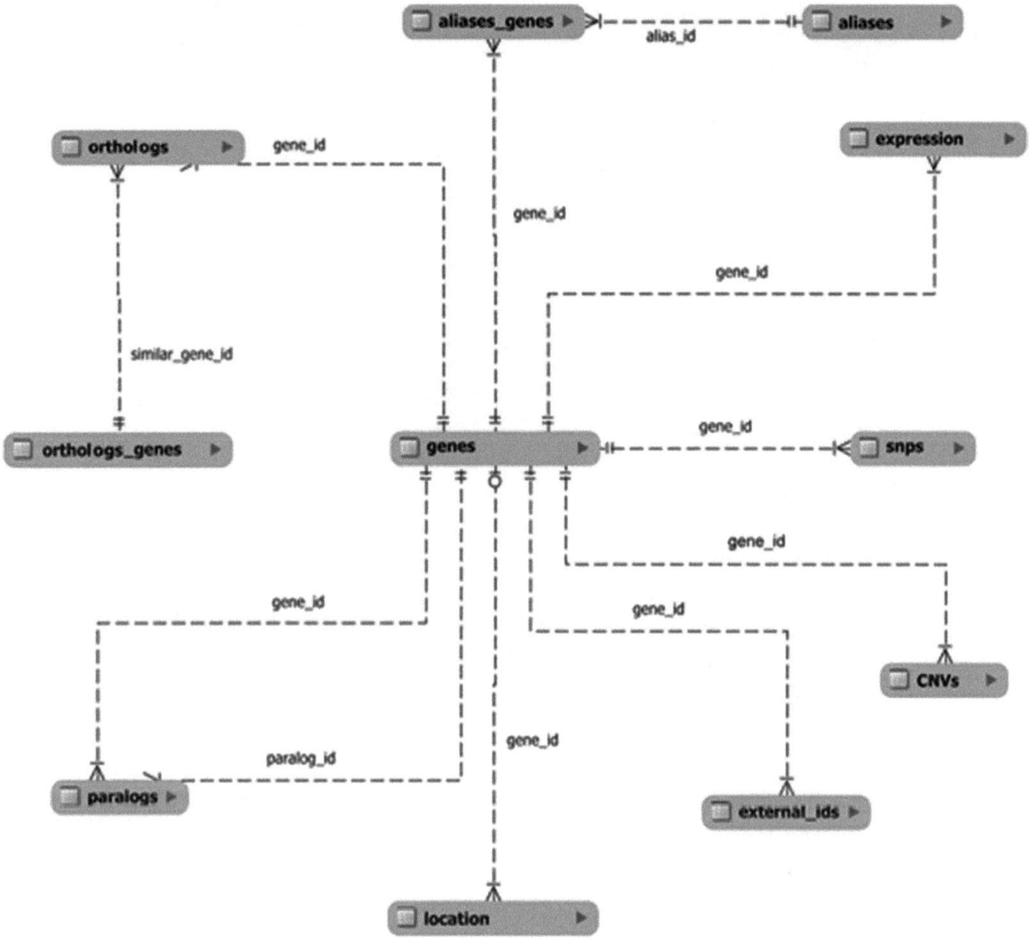

Fig. 2 HORDE version #43 design in a relational model. The *boxes* represent the database tables. The *dashed arrows*, labeled with the foreign key fields, show the relationships between the tables

Table 1
Summary of the OR repertoires listed in HORDE

Species	Reference	Number of intact ORs	Number of pseudogenes
Human	(13)	390 (387)	468 (415)
Chimpanzee	(16)	397 (380)	693 (414)
Dog	(33)	859	315
Opossum	(33)	613	898
Platypus	(15)	350	404

Also given in parenthesis, the gene count of the human and chimpanzee OR repertoires from the study of (65), for comparison (see Note 4)

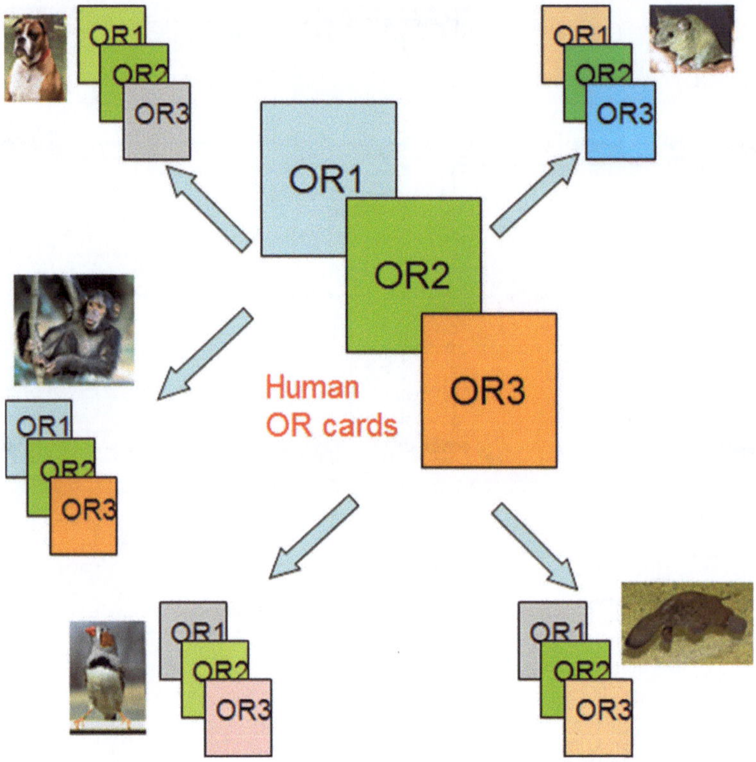

Fig. 3 HORDE is a human-centric database but includes OR repertoires from other vertebrates. The information is presented to the user via an "OR card" for all species. The information for human ORs is extensive: it is integrated with external relevant sources. The human OR card contains links to the ortholog OR cards in other species

Based on the classification, HORDE provides a HUGO-approved systematic nomenclature (http://www.gene.ucl.ac.uk/nomenclature/genefamily.shtml) that affords an instant guide to the position of a gene in a phylogenetic tree (36). The nomenclature system consists of a root "OR," followed by the family numeral, the subfamily letter(s), and a numeral representing the individual gene within the subfamily. For example, OR3A1 is an OR gene of family 3, subfamily A, and OR7E12P is an OR pseudogene of family 7 subfamily E. This nomenclature system for human ORs is now widely.

3.2 The Human OR Card

A key feature in HORDE is a "card" for every human OR gene (Figs. 3 and 4a), a concept adopted from the GeneCards compendium of human genes (9, 37). The OR card includes rich annotation relevant to this gene, including links to internal tools as well as to appropriate external databases. A description of the different sections of the human OR card is provided below.

3.2.1 Gene Nomenclature and Phylogeny

This section contains information regarding the classification of the OR, its aliases and its functionality (being intact or pseudogene). OR genes with in-frame stop codons or frame shifts are

a

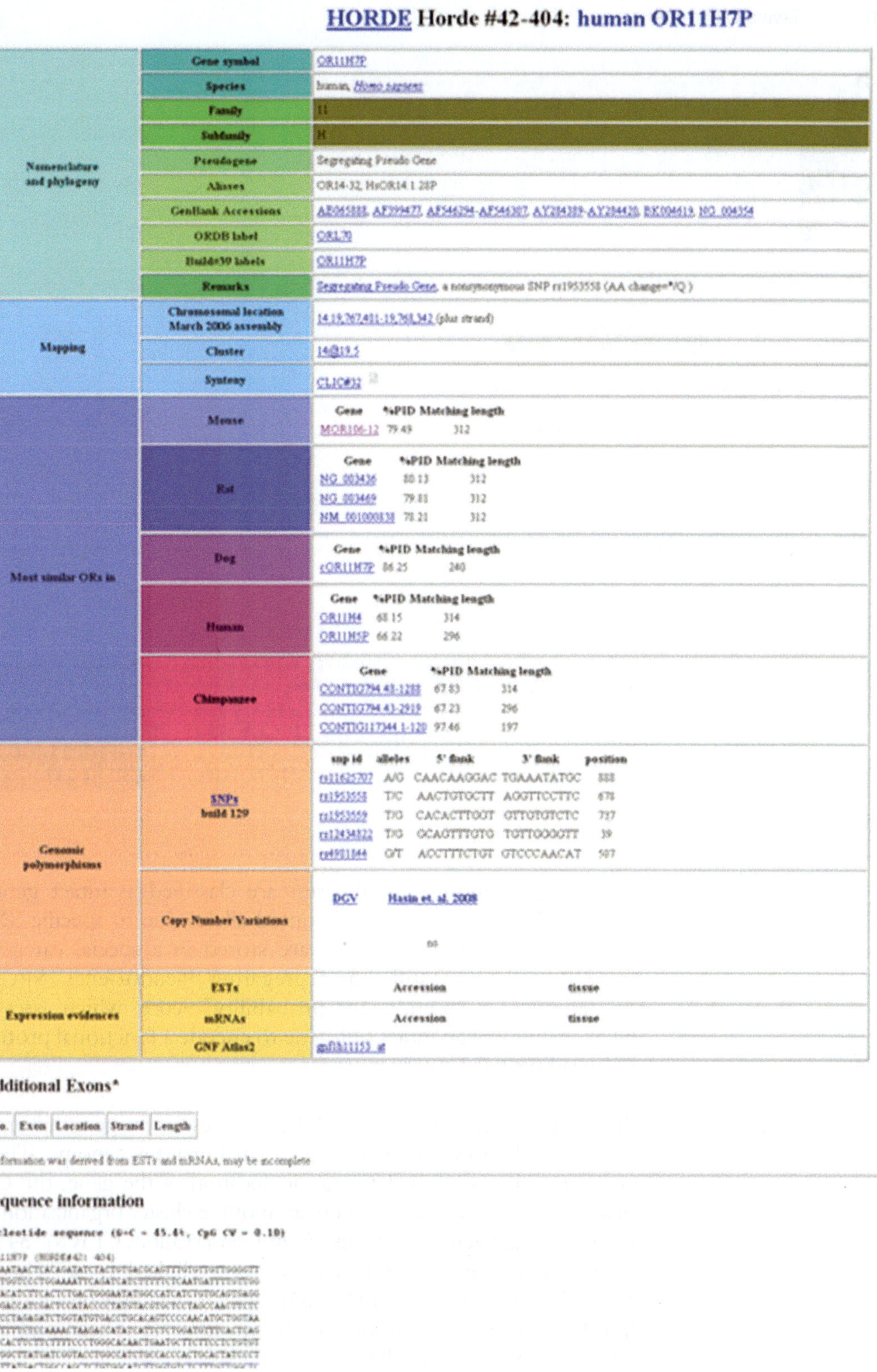

Fig. 4 (**a**) An example of human OR card. Note, that the OR presented, OR11H7P, is a Segregating PseudoGene, due to a SNP, which alters a frame disrupting stop codon into glutamine. A recent genomic study reported on an association between this SNP and isovaleric acid sensitivity (60). (**b**) An example of nonhuman OR card. The canine's ortholog of OR11H7P

b

Dog olfactory receptor gene (cOR11H7)

Gene symbol	cOR11H7
Symbol in Version_1	cOR11H7P
Family	11
Subfamily	H
Pseudo	no
Location	chr15:20594791-20595786
Strand	*plus*
Synteny	CLIC #32

Sequence information

Nucleotide sequence (G+C = 45.4%, CpG CV = 0.21)

>cOR11H7
ATGTGTCAGCAGCAAAATGACTTGTCACTGTCTTCTCTCCACAGTTTGGAGACCATGAATAAGTCAGAGA
TATCTACTGTGACACAGTTTGTCTTGTTGGGTTTTCCTGGTCCCTGGAAAATACAGATCATGCTTTTCTC
AATGATTTTGTTGGTCTATATCTTGACTCTAACTGGGAATACGGCCATCATTTGTGCAGTGAGATGGGAC
CATCGACTCCATACCCCTATGTATGTGCTCCTGGCAAACTTCTCCTTCCTAGAAATCTGGTATGTGACTT
GCACCATTCCCAACATGCTGGTCAATTTACTTTCCAAAACAAAGACCATATCGTTCTCTGGTTGCTTCGC
TCAGTTCTACTTCTTCTTTTCCCTGGGCACAACTGAGTGCTTCTTCCTCTGTGTCATGGCGTATGATAGG

Fig. 4 (continued)

classified as pseudogenes; the rest are classified as intact genes. Genes with both intact and disrupted alleles due to specific SNP that leads to allele inactivation are stored in a special category are (38), and are classified as *S*egregating *P*seudo*G*enes (SPGs). We determine a "pseudogene probability" score, which assesses the probability of an intact OR gene to encode a functional protein (Subheading 4.3.1).

3.2.2 Mapping

The mapping section provides a global overview of the genomic organization of the OR superfamily in the mammalians genome. Thus, information is given on the genomic location of the gene, the OR cluster (Fig. 5) and on the conservation of the cluster organization in five mammalian genomes (*CL*usters *I*n *C*onservation, CLICs) (34).

The information about the cluster is linked to a "cluster viewer," a special feature of HORDE, which facilitates browsing of the ORs organization along human chromosomes (Subheading 4.3.3 provides a bit more details).

CLICs information is linked to a dedicated Web page within HORDE (http://genome.weizmann.ac.il/horde/CLICs/), which presents the syntenic clusters in other species (Subheading 4.3.2).

3.2.3 OR Paralogs and Orthologs

The OR superfamily have been subjected to frequent gene recombination, gene conversion, duplication, and translocation events (14, 39). These events make the identification of orthologs and paralogs a nontrivial task. By definition, HORDE does not attempt to identify orthologs and paralogs authoritatively, but rather supplies for the documented gene (probe) the top BLASTP hits in the human repertoire and in some other mammals (Fig. 6). If the best hit in a given species has X % identity to the probe, HORDE presents all results that have identity scores >(X − 5) % to the probe.

3.2.4 Genomic Polymorphisms

Human ORs are enriched in genomic variations (40, 41, 66), a result of gene redundancy that also leads to the gradual evolutionary diminution of the human OR repertoire. HORDE aims to contain a comprehensive collection of genomic variations in the OR genes, presented in the genomic polymorphisms section.

SNPs and small indels are extracted from dbSNP (42), the 1,000 genomes project (http://www.1000genomes.org/page.php), and an in-house next generation sequencing of 100 genes. CNVs are extracted from the Database of Genomic Variants (DGV, http://projects.tcag.ca/variation/) (43), MoDIL (http://compbio.cs.toronto.edu/modil/ (44)) and dbSNP (build 131), as well as from a published paper (45).

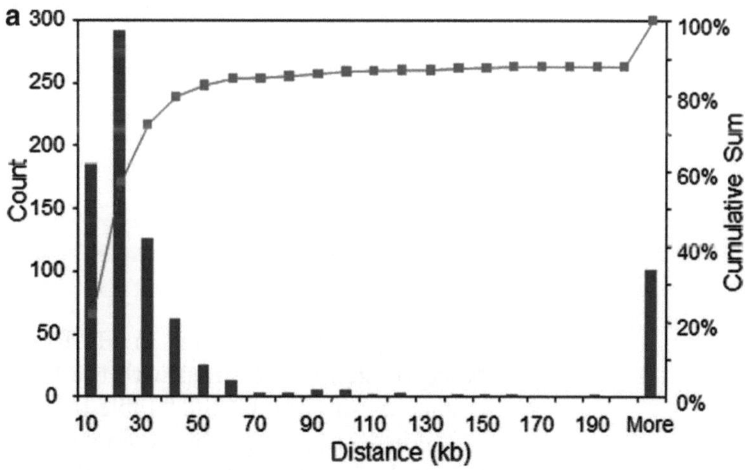

Fig. 5 Genomic organization of the human OR repertoire. (**a**) ORs are disposed in the genome in a compact clusters, with a mean distance of less than 20 ± 15 kb. The *bars* show the distribution of the distances of each two consecutive ORs. The cumulative sum is represented by a *line dotted with squares*. (**b**) HORDE divides the human OR repertoire into 136 clusters ranging from one (single ton) to 97 ORs per cluster. The largest 20 clusters are shown, including their OR families composition

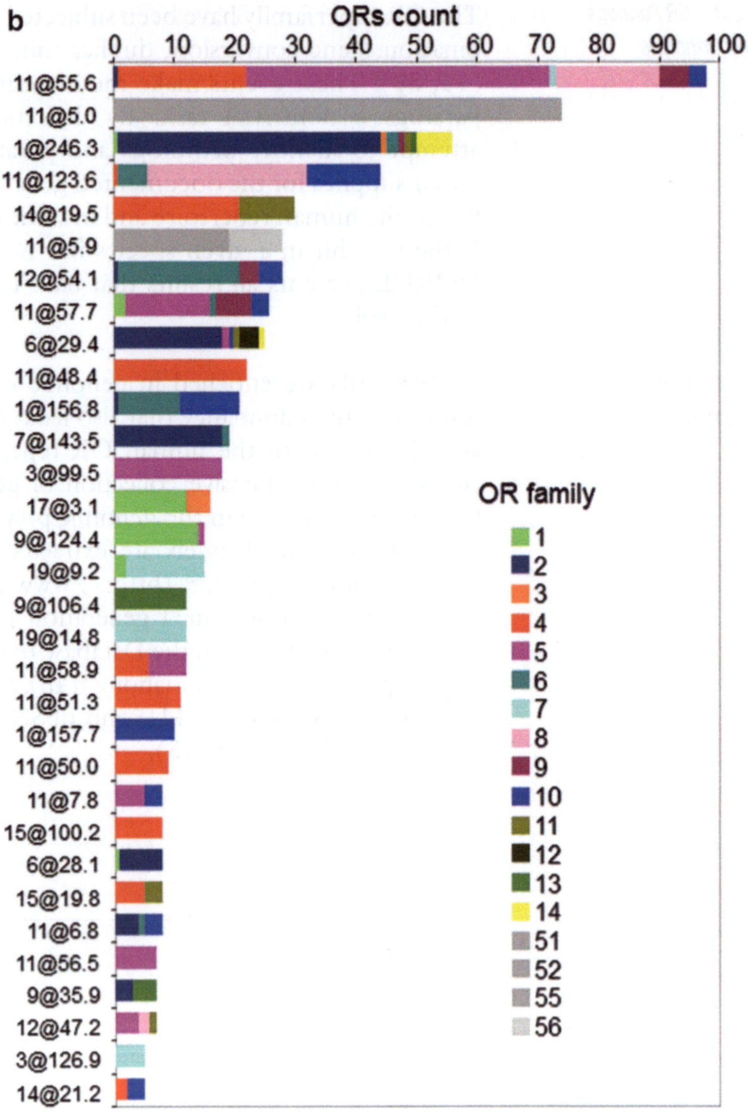

Fig. 5 (continued)

All genomic variations in HORDE (Table 2) are collected based on genomic overlap with the OR coordinates. SNP data is extracted using the UCSC "Table browser" tool (http://genome.ucsc.edu/cgi-bin/hgTables?command=start) and further analyzed by our Perl script that interprets their effect on the protein (i.e., whether they are synonymous or non-synonymous). SNPs with a *map weight*>1 (a parameter indicating nonunique genome mapping (46)), are shown with multiple map targets. (see Note 3)

3.2.5 Expression Evidence and Additional Exons

ORs are expressed in the olfactory epithelium, but also ectopically in many other tissues (24–26). The expression evidence section provides information of gene expression extracted from BioGPS

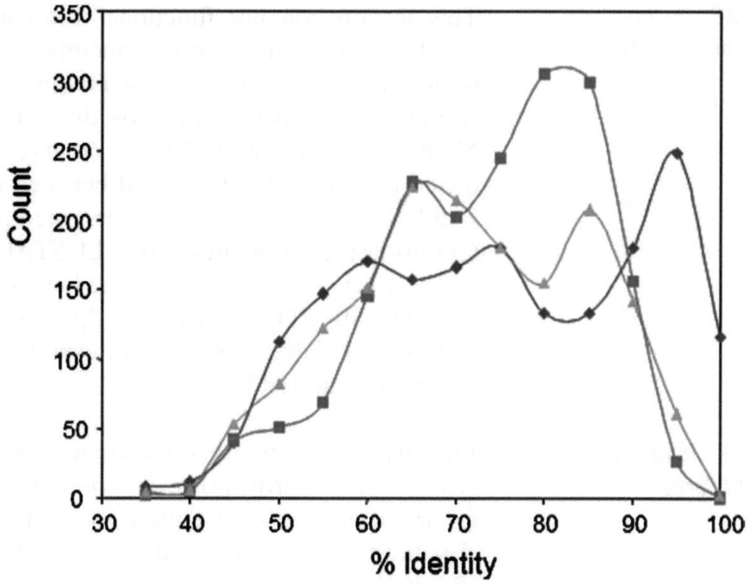

Fig. 6 A distribution of the % identity values of the human most similar ORs in mouse (*line with squares*), dog (*line with triangles*) and human (*line with diamonds*). Such a distribution can be used to study orthology and paralogy relationship

Table 2
A summary of all genomic polymorphic events in HORDE (66)

ORs	SNPs	In/Dels (<1,000 bp)	CNVs	CNVs with deletion allele
Intact	5,362	204	291	98
Pseudo	5,591	452	360	133

In/Dels = insertions or deletions

DNA arrays (47, 48) and *in silico* data mining of ESTs and mRNA (12, 25). The full details of the data mining and analysis of ESTs and mRNAs are described in (25) and (12).

The mRNA that encodes ORs typically includes a single downstream exon that contains the entire ~900 bp coding region, and encompasses ~1–2 kb 3′-UTR. The upstream part of the transcript often has several short 5′ untranslated exons, with evidence for alternative splicing, both by standard intron removal and by in-exon splicing (49–52). Unfortunately, the untranslated regions of the OR genes are not conserved and cannot be readily identified by current gene prediction programs or by HORDE's pipeline. By analyzing spliced ESTs, we defined and added to the database information on 388 untranslated exons belonging to 149 ORs (12).

3.2.6 Protein Domains and Features

This section contains functional annotation of the OR protein which includes the seven transmembrane domains, positions suspected as the putative odorant binding site residues, termed the complementary determining residues (CDRs) (53), a conserved N-glycosylation site, and two conserved cysteines that are predicted to form a disulfide bond between extracellular loops 1 and 2 (53, 54). The annotation is performed by aligning each OR protein using the profile mode of CLUSTALW to the highly curated and annotated alignment published in (53).

Currently (build #43), the "protein domains and features" section is available to the users through a link from the main OR card to a dedicated html page.

3.3 Nonhuman OR Cards

Information about other vertebrate ORs is also presented by a "card." These cards, however, contain less information than that for humans. The card typically includes the following: the OR classification, genomic location, and sequences (Fig. 4b). The dog and opossum OR cards contain links to their relevant CLIC.

4 HORDE Query Tools

HORDE is equipped with several search tools, allowing a text-free search as well as sequence search.

4.1 Sequence Search Tool

The BLAST sequence search tool provides access to a large collection of OR repertoires that we have included within HORDE. Besides the OR repertoires listed in Table 1, library access is provided to the OR repertoires of the mouse (31, 55), rat (56) and chicken (extracted from GenBank). Another library, named the "zoo library," contains 2,235 OR partial sequences we have mined from 47 vertebrates collected from GenBank.

4.2 Text Search Tools

HORDE offers the user two text-search tools. The first is a "free text" tool. For example, a search with the text "Segregating PseudoGene" will result in a list of all SPGs in the database. A search using the term "rs379147," will return a link to OR1E1, the gene that contains this SNP. The second tool allows the user to select for a group of ORs based on a shared chromosomal location, or classification. For example, under the option *Select all ORs from*: inserting the number "1" for the *family*, results in a list of all ORs from family 1. The nucleotide or protein sequence can be then retrieved for all or selected ORs in this list.

4.3 HORDE Analysis Tools

4.3.1 CORP (http://genome.weizmann.ac.il/horde/CORP/)

The probabilistic Classifier for Olfactory Receptor Pseudogenes (CORP) algorithm (57) computes the probability of an OR gene with intact open reading frame to encode a nonfunctional protein (i.e., a pseudogene) by examining the deviation of its protein sequences from the OR functionally crucial consensus. It accepts as an input an OR protein sequence and returns the pseudogene probability score of the gene.

4.3.2 CLICs (http://genome.weizmann.ac.il/horde/CLICs/)

CLIC (Cluster In Conservation) is a tool for genome-wide definition of genomic gene clusters conserved in multiple species (34). The CLIC algorithm was applied to the ORs of five mammals. 48 multispecies CLICs were discovered, covering more than 90 % of all OR genes. The CLIC Web page contains a list of all CLICs and full details of the ORs participating in each CLIC from all compared species.

4.3.3 The Cluster Viewer

This presents the cluster details (location, size, and OR family composition) and the OR cluster members. A link to the closest OR clusters allows the user to browse the OR clusters a long a given chromosome. A cluster is a group of ORs with maximal intergenic distance of a 100 kb (Fig. 5).

5 Notes

1. Although HORDE data-mining routine was successfully applied to several vertebrate genomes, it might be argued that the use of BLAST search is too conservative and the result is biased toward genes showing similarity to the "OR baits," potentially missing some *bona fide* OR genes. For this reason, we make sure that we perform the data mining procedure until known GPCR proteins are identified.
2. HORDE is currently stored in a flat-file format equipped with Perl-CGI interface. For the next version (August 2012), all the data stored into HORDE will be ported into a relational database. Leveraging the flexibility and complexity that such a data-infrastructure can contribute, it will allow complex user-queries and integration with other sources.
3. The information on the complete OR gene strecture (Subheading 3.2.5) is partial and preliminary. A new mining procedure is now under development and will be based on the AceView analysis (58), which performed the same analysis but on a whole genome scale and thus is more accurate.
4. The increasing number of sequenced individual genomes will dramatically increase our knowledge of the known genomic variations in the human ORs. Of special relevance are deleterious

variations, nonsense SNPs, which cause premature stop codon, null allele CNVs (deletions) leading to OR inactivation (45, 59), or substitution in a highly conserved amino acid. Such polymorphisms might lead to phenotypic differences in olfactory acuity and perception, as was demonstrated in the study of (60) and (61). One of HORDE's challenges is to present a continuously updated compendium of genomic variations, thus providing the infrastructure for future genomic genetic association studies of OR genotypes to olfactory phenotypes.

5. A companion database is now available for olfactory auxiliary genes related to transduction and sensory neuronal integrity (http://genome.weizmann.ac.il/GOSdb/, REF: PMID: 22936402).

Acknowledgements

This research was supported by the Israel Science Foundation Heritage fund, LifeMap Sciences Inc. California (USA), as well as the Crown Human Genome Center and the Nella and Leon Benoziyo Center for Neurological Diseases at the Weizmann Institute of Science.

References

1. Touhara K, Vosshall LB (2009) Sensing odorants and pheromones with chemosensory receptors. Annu Rev Physiol 71:307–332
2. Kaupp UB (2010) Olfactory signalling in vertebrates and insects: differences and commonalities. Nat Rev Neurosci 11:188–200
3. Rodriguez-Gil DJ, Treloar HB, Miller A et al (2009) Onset of odorant receptors. Ann N Y Acad Sci 1170:18–20
4. Nei M, Niimura Y, Nozawa M (2008) The evolution of animal chemosensory receptor gene repertoires: roles of chance and necessity. Nat Rev Genet 9:951–963
5. Rouquier S, Giorgi D (2007) Olfactory receptor gene repertoires in mammals. Mutat Res 616:95–102
6. Kambere MB, Lane RP (2007) Co-regulation of a large and rapidly evolving repertoire of odorant receptor genes. BMC Neurosci 8(Suppl 3):S2
7. Crasto C, Marenco L, Miller P et al (2002) Olfactory receptor database: a metadata-driven automated population from sources of gene and protein sequences. Nucleic Acids Res 30:354–360
8. Horn F, Bettler E, Oliveira L et al (2003) GPCRDB information system for G protein-coupled receptors. Nucleic Acids Res 31: 294–297
9. Safran M, Chalifa-Caspi V, Shmueli O et al (2003) Human gene-centric databases at the Weizmann Institute of Science: GeneCards, UDB, CroW 21 and HORDE. Nucleic Acids Res 31:142–146
10. Glusman G, Yanai I, Rubin I et al (2001) The complete human olfactory subgenome. Genome Res 11:685–702
11. Fuchs T, Glusman G, Horn-Saban S et al (2001) The human olfactory subgenome: from sequence to structure and evolution. Hum Genet 108:1–13
12. Olender T, Feldmesser E, Atarot T et al (2004) The olfactory receptor universe–from whole genome analysis to structure and evolution. Genet Mol Res 3:545–553
13. Olender T, Lancet D, Nebert DW (2008) Update on the olfactory receptor (OR) gene superfamily. Hum Genomics 3:87–97
14. Niimura Y (2009) Evolutionary dynamics of olfactory receptor genes in chordates: interaction between environments and genomic contents. Hum Genomics 4:107–118
15. Warren WC, Hillier LW, Marshall Graves JA et al (2008) Genome analysis of the platypus

reveals unique signatures of evolution. Nature 453:175–183
16. Gilad Y, Man O, Glusman G (2005) A comparison of the human and chimpanzee olfactory receptor gene repertoires. Genome Res 15:224–230
17. Kishida T, Kubota S, Shirayama Y et al (2007) The olfactory receptor gene repertoires in secondary-adapted marine vertebrates: evidence for reduction of the functional proportions in cetaceans. Biol Lett 3:428–430
18. Karro JE, Yan Y, Zheng D et al (2007) Pseudogene.org: a comprehensive database and comparison platform for pseudogene annotation. Nucleic Acids Res 35:D55–D60
19. Mombaerts P (2004) Genes and ligands for odorant, vomeronasal and taste receptors. Nat Rev Neurosci 5:263–278
20. Hasin-Brumshtein Y, Lancet D, Olender T (2009) Human olfaction: from genomic variation to phenotypic diversity. Trends Genet 25: 178–184
21. Menashe I, Lancet D (2006) Variations in the human olfactory receptor pathway. Cell Mol Life Sci 63:1485–1493
22. Nguyen MQ, Zhou Z, Marks CA et al (2007) Prominent roles for odorant receptor coding sequences in allelic exclusion. Cell 131: 1009–1017
23. Imai T, Sakano H (2009) Odorant receptor gene choice and axonal projection in the mouse olfactory system. Results Probl Cell Differ 47:57–75
24. De la Cruz O, Blekhman R, Zhang X et al (2009) A signature of evolutionary constraint on a subset of ectopically expressed olfactory receptor genes. Mol Biol Evol 26:491–494
25. Feldmesser E, Olender T, Khen M et al (2006) Widespread ectopic expression of olfactory receptor genes. BMC Genomics 7:121
26. Zhang X, De la Cruz O, Pinto JM et al (2007) Characterizing the expression of the human olfactory receptor gene family using a novel DNA microarray. Genome Biol 8:R86
27. Fukuda N, Yomogida K, Okabe M et al (2004) Functional characterization of a mouse testicular olfactory receptor and its role in chemosensing and in regulation of sperm motility. J Cell Sci 117:5835–5845
28. Xu LL, Stackhouse BG, Florence K et al (2000) PSGR, a novel prostate-specific gene with homology to a G protein-coupled receptor, is overexpressed in prostate cancer. Cancer Res 60:6568–6572
29. Zhang X, Firestein S (2009) Genomics of olfactory receptors. Results Probl Cell Differ 47:25–36
30. Young JM, Trask BJ (2002) The sense of smell: genomics of vertebrate odorant receptors. Hum Mol Genet 11:1153–1160
31. Young JM, Friedman C, Williams EM et al (2002) Different evolutionary processes shaped the mouse and human olfactory receptor gene families. Hum Mol Genet 11:535–546
32. Niimura Y, Nei M (2005) Evolutionary dynamics of olfactory receptor genes in fishes and tetrapods. Proc Natl Acad Sci USA 102:6039–6044
33. Olender T, Fuchs T, Linhart C et al (2004) The canine olfactory subgenome. Genomics 83:361–372
34. Aloni R, Olender T, Lancet D (2006) Ancient genomic architecture for mammalian olfactory receptor clusters. Genome Biol 7:R88
35. Warren WC, Clayton DF, Ellegren H et al (2010) The genome of a songbird. Nature 464:757–762
36. Glusman G, Bahar A, Sharon D et al (2000) The olfactory receptor gene superfamily: data mining, classification and nomenclature. Mamm Genome 11:1016–1023
37. Safran M, Dalah I, Alexander J et al (2010) GeneCards Version 3: the human gene integrator. Database (Oxford) 2010:baq020
38. Menashe I, Man O, Lancet D et al (2003) Different noses for different people. Nat Genet 34:143–144
39. Lapidot M, Pilpel Y, Gilad Y et al (2001) Mouse-human orthology relationships in an olfactory receptor gene cluster. Genomics 71:296–306
40. Alonso S, Lopez S, Izagirre N et al (2008) Overdominance in the human genome and olfactory receptor activity. Mol Biol Evol 25: 997–1001
41. Nozawa M, Kawahara Y, Nei M (2007) Genomic drift and copy number variation of sensory receptor genes in humans. Proc Natl Acad Sci 104:20421 20426
42. Day IN (2010) dbSNP in the detail and copy number complexities. Hum Mutat 31:2–4
43. Zhang J, Feuk L, Duggan GE et al (2006) Development of bioinformatics resources for display and analysis of copy number and other structural variants in the human genome. Cytogenet Genome Res 115:205–214
44. Lee S, Hormozdiari F, Alkan C et al (2009) MoDIL: detecting small indels from clone-end sequencing with mixtures of distributions. Nat Methods 6:473–474
45. Hasin Y, Olender T, Khen M et al (2008) High-resolution copy-number variation map reflects human olfactory receptor diversity and evolution. PLoS Genet 4:e1000249

46. http://www.ncbi.nlm.nih.gov/bookshelf/br.fcgi?book=helpsnpfaq&part=Build.The_dbSNP_Mapping_Process#Build.Mapping_Weight
47. Wu C, Orozco C, Boyer J et al (2009) BioGPS: an extensible and customizable portal for querying and organizing gene annotation resources. Genome Biol 10:R130
48. Su AI, Wiltshire T, Batalov S et al (2004) A gene atlas of the mouse and human protein-encoding transcriptomes. Proc Natl Acad Sci USA 101:6062–6067
49. Fleischer J, Breer H, Strotmann J (2009) Mammalian olfactory receptors. Front Cell Neurosci 3:9
50. Young JM, Shykind BM, Lane RP et al (2003) Odorant receptor expressed sequence tags demonstrate olfactory expression of over 400 genes, extensive alternate splicing and unequal expression levels. Genome Biol 4:R71
51. Michaloski JS, Galante PA, Malnic B (2006) Identification of potential regulatory motifs in odorant receptor genes by analysis of promoter sequences. Genome Res 16:1091–1098
52. Sosinsky A, Glusman G, Lancet D (2000) The genomic structure of human olfactory receptor genes. Genomics 70:49–61
53. Man O, Gilad Y, Lancet D (2004) Prediction of the odorant binding site of olfactory receptor proteins by human-mouse comparisons. Protein Sci 13:240–254
54. Menon ST, Han M, Sakmar TP (2001) Rhodopsin: structural basis of molecular physiology. Physiol Rev 81:1659–1688
55. Zhang X, Firestein S (2002) The olfactory receptor gene superfamily of the mouse. Nat Neurosci 5:124–133
56. Gibbs RA, Weinstock GM, Metzker ML et al (2004) Genome sequence of the Brown Norway rat yields insights into mammalian evolution. Nature 428:493–521
57. Menashe I, Aloni R, Lancet D (2006) A probabilistic classifier for olfactory receptor pseudogenes. BMC Bioinformatics 7:393
58. Thierry-Mieg D, Thierry-Mieg J (2006) AceView: a comprehensive cDNA-supported gene and transcripts annotation. Genome Biol 7(Suppl 1):S12 1–S12 14
59. Young JM, Endicott RM, Parghi SS et al (2008) Extensive copy-number variation of the human olfactory receptor gene family. Am J Hum Genet 83:228–242
60. Menashe I, Abaffy T, Hasin Y et al (2007) Genetic elucidation of human hyperosmia to isovaleric acid. PLoS Biol 5:e284
61. Keller A, Zhuang H, Chi Q et al (2007) Genetic variation in a human odorant receptor alters odour perception. Nature 449:468–472
62. Kent WJ (2002) BLAT–the BLAST-like alignment tool. Genome Res 12:656–664
63. Camacho C, Coulouris G, Avagyan V et al (2009) BLAST+: architecture and applications. BMC Bioinformatics 10:421
64. Pearson WR, Wood T, Zhang Z et al (1997) Comparison of DNA sequences with protein sequences. Genomics 46:24–36
65. Go Y, Niimura Y (2008) Similar numbers but different repertoires of olfactory receptor genes in humans and chimpanzees. Mol Biol Evol 25:1897–1907
66. Olender T, Waszak SM, Viavant M et al (2012) Personal receptor repertoires: olfaction as a model.BMC Genomics. Aug 21;13:414. doi: 10.1186/1471-2164-13-414

Chapter 3

Identification of Olfactory Receptor Genes from Mammalian Genome Sequences

Yoshihito Niimura

Abstract

Olfaction is essential for the survival of mammals. Diverse odorant molecules in the environment are detected by olfactory receptors (ORs) expressed in the olfactory epithelium of the nasal cavity. In general, mammalian genomes harbor ~1,000 OR genes, which form the largest multigene family in mammals. The recent advances in genome sequencing technology have enabled us to computationally identify nearly complete repertoires of OR genes from various organisms. Such studies have revealed that the numbers of OR genes are highly variable among organisms depending on their living environments.

Because OR genes are intronless, it is possible to find all OR genes by conducting homology searches against the genome sequences using known OR genes as queries. However, some caution is necessary during the process of extracting intact coding sequences of OR genes and distinguishing among OR and non-OR genes. Presented here is a description of bioinformatics methods to identify the entire OR gene repertoires from mammalian genome sequences.

Key words Olfactory receptor, Multigene family, Bioinformatics, Mammalian genome, G-protein coupled receptor

1 Introduction

Diverse odor molecules in the environment are detected by olfactory receptors (ORs) that are expressed in the olfactory epithelium of the nasal cavity. To deal with enormous diversity of odorants, the mammalian genomes harbor OR genes with varied sequences. The genomes of many mammalian species contain ~1,000 or more OR genes, which form the largest multigene family in mammals (*for reviews*, see (1, 2)). OR genes were first identified from rats by Linda Buck and Richard Axel in 1991 (3), a discovery that led to them being awarded the Nobel Prize in 2004.

An OR is a G-protein coupled receptor (GPCR) with seven α-helical transmembrane (TM) regions. ORs are, on average, approximately, 310 amino acids long. OR genes belong to rhodopsin-like

Chiquito J. Crasto (ed.), *Olfactory Receptors: Methods and Protocols*, Methods in Molecular Biology, vol. 1003,
DOI 10.1007/978-1-62703-377-0_3,

GPCR superfamilies, which include opsins for detecting light and receptors for various ligands such as neurotransmitters, peptide hormones, chemokines, lipids, and nucleotides, and others (4). OR genes do not have any introns in their coding regions. Introns are often present in the 5′ untranslated regions of an OR gene; however, different mRNA isoforms generated by alternative splicing of noncoding exons result in the same protein (5).

OR genes are present in all vertebrate species. Fish have much smaller repertoires of OR genes (~100) than mammals. However, although OR gene repertoires in fish are small, their OR genes are more diverse in sequence than those of mammals (6, 7). Mammalian, reptilian, and avian OR genes can be clearly classified into two groups, Class I and Class II (8), on the basis of their amino acid sequence similarities. On the other hand, based on sequence similarities, amphibians and fish have additional groups of OR genes (6, 7).

Amphioxus, the most basal chordate species, also possesses vertebrate-like OR genes (7). Therefore, the origin of OR genes can be traced back to the common ancestor of all chordate species. Insects and nematodes also have OR genes. However, those in chordates, insects, and nematodes do not show sequence similarities, suggesting multiple origins of OR genes during animal evolution (1, 2, 9).

Thanks to the recent advances in genome sequencing technologies, whole genome sequences of various organisms have become available. We have established computational methods to identify the entire set of OR genes in a given species, and have conducted comparative analyses of the OR gene repertoires of various organisms (6, 7, 10–15). These studies have showed that the numbers of OR genes are highly variable among different species. For example, higher primates including humans generally have smaller repertoires of (functional) OR genes (<400) than most other mammals (10–15). This observation would reflect the fact that higher primates rely on vision rather than olfaction, and thus their olfactory ability has retrogressed. Another feature of mammalian OR gene families is a high fraction (20–60 %) of pseudogenes in each species (13, 15). Moreover, evolutionary analyses revealed that the numbers of OR genes have dynamically changed during evolution due to frequent gene duplications and pseudogenization events (6, 12, 13, 15).

Here are described details of the methods to identify OR genes from mammalian genomes. These methods are also applicable and extensible to reptiles and birds, because these species have the same groups of OR genes as mammals. However, to identify OR genes from the genomes of amphibians or fish, minor methodological modifications are necessary because of the higher sequence diversity in their OR genes compared with those of mammals.

2 Materials

1. The genome sequences of various mammalian species can be downloaded from the Web sites of the University of California Santa Cruz (UCSC; genome.ucsc.edu), Ensembl Genome Browser (ensembl.org), and the Genome Sequencing Center at Washington University School of Medicine (genome.wustl.edu), the Broad Institute (www.broadinstitute.org), and others.
2. The latest version of mammalian OR gene repertoires, which are used as queries for the TBLASTN searches, can be obtained from Niimura and Nei (13), Go and Niimura (14), and Matsui et al. (15). Other databases such as GenBank (www.ncbi.nlm.nih.gov/genbank) and Olfactory Receptor Database (ORDB; senselab.med.yale.edu/ordb) (see Chapter 1) (16) also contain many OR gene sequences.

3 Methods

The strategy to identify OR genes from a given genome sequence is illustrated in Fig. 1. Here are classified OR genes identified from the genome sequences into three categories, functional genes, truncated genes, and pseudogenes. A functional gene is defined as an intact sequence that potentially encodes a functional OR, while a pseudogene is defined as a sequence containing interrupting stop codons, frameshifts, and/or gaps within conserved regions. A truncated gene is a partial intact sequence containing a part of OR gene sequence and is located at the contig end.

A truncated gene is either a functional gene or a pseudogene. If the genome sequencing is completed, it can be classified as either. For a low-coverage genome, the lengths of contigs tend to be short due to incomplete assembly. Thus, the number of truncated genes becomes large. The distinction between truncated genes and pseudogenes is necessary for estimating the fraction of pseudogenes in a given species, because the number of functional OR genes is underestimated if only intact genes are regarded as functional (13, 15).

3.1 Identification of Functional OR Genes

1. The first process to identify OR genes is to conduct TBLASTN (17) searches against a given genome sequence using known OR genes as queries. For query sequences, use a variety of known OR genes (see Note 1). The *E*-value should be 1e−10 for the identification mammalian genes as ORs.
2. Because multiple sequences are used as queries, many query sequences may hit the same genomic region. The genomic

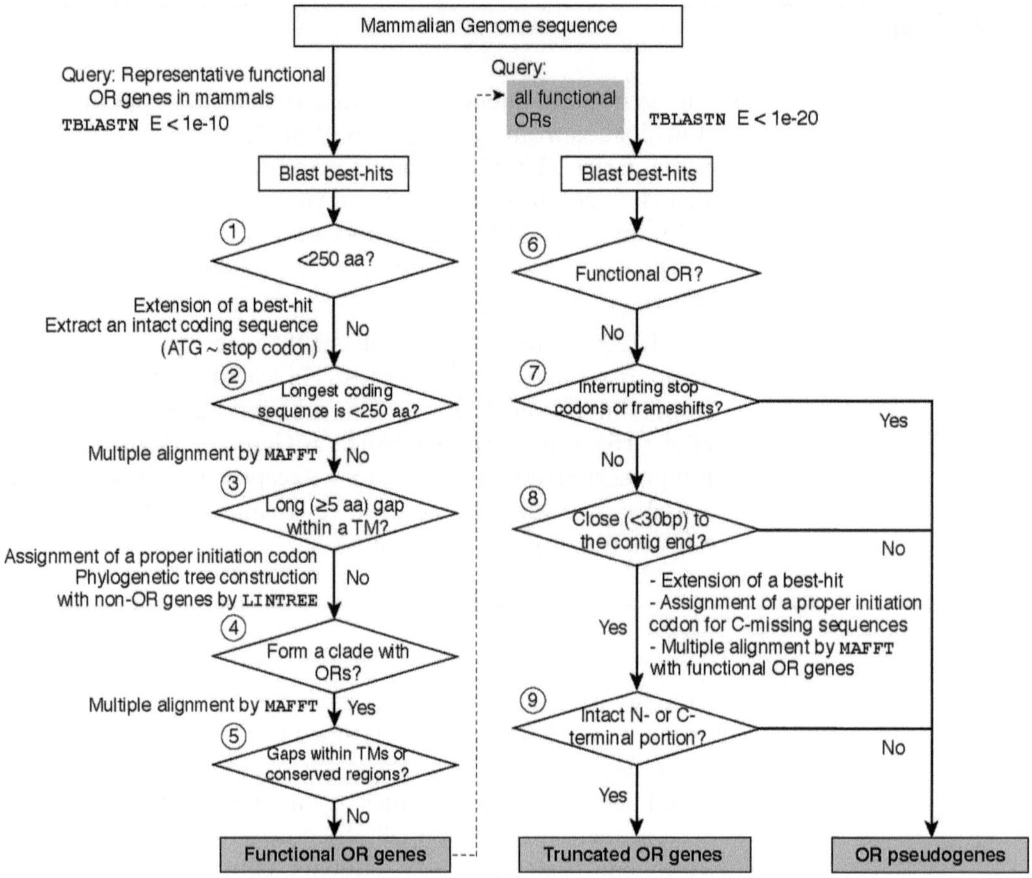

Fig. 1 Flowchart for the identification of functional OR genes, truncated OR genes, and OR pseudogenes from mammalian genome sequences

sequence that corresponds to a query showing the lowest *E*-value is the "best-hit" (Fig. 2a). Extract all the best-hit sequences from the genome.

3. Functional OR genes are identified from the best-hit sequences obtained above. To exclude nonfunctional OR genes or non-OR genes from the best-hits, the following criteria are used. First, discard the best-hit sequences that are shorter than 250 amino acids (criterion 1 in Fig. 1; see Note 2). Sometimes a query hits a genomic region that is separated into two or more fragments due to frameshifts. In such a case, examine the longest sequence among the fragments. Discard the best-hit if its length is less than 250 amino acids.

4. For each of the best-hit sequences, extend it in both directions along the genome sequence. Then, extract the longest intact coding sequence starting from an ATG codon and ending with a stop codon (TAA, TAG, or TGA) without any interrupting stop codons (Fig. 2b). If the length of the intact coding

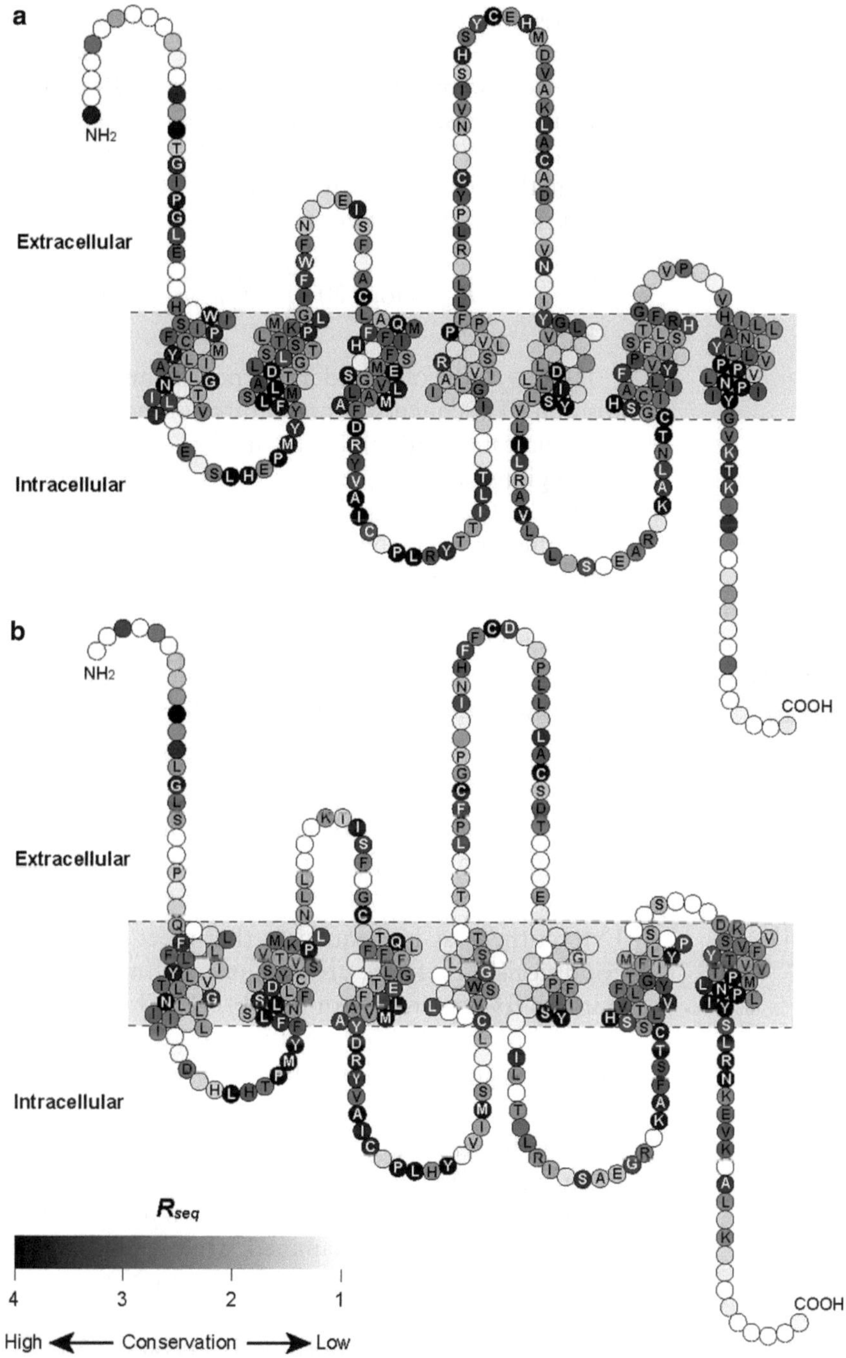

Fig. 4 Conservative amino acid sites for Class I (**a**) and Class II (**b**) OR genes. The *darkness* of a *circle* represents an extent of amino acid conservation (R_{seq}) at each position. At the amino acid site with the R_{seq} value of 1.5 or larger, the most frequent amino acid at the position is shown in the *single letter code*

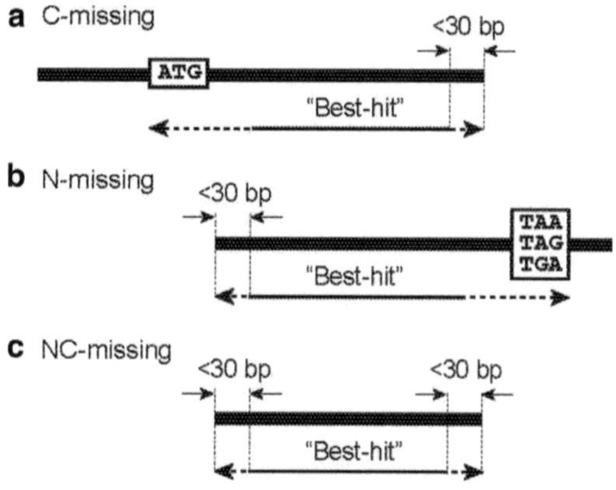

Fig. 5 Extension of a best-hit for C-missing (**a**), N-missing (**b**), and NC-missing sequences (**c**)

most downstream in-frame codon. As for the ATG codon, take the most upstream in-frame one without any interrupting stop codons (Fig. 5a).

7. For an N-missing sequence, extend it to take a sequence from the most upstream in-frame codon to the stop codon (Fig. 5b).
8. For an NC-missing sequence, extend it to take the longest sequence from the most upstream in-frame codon to the most downstream one (Fig. 5c).
9. Construct a multiple alignment using the extended N-missing, C-missing, and NC-missing sequences together with functional OR genes identified above by the program E INS-i in MAFFT (18).
10. For each of the N-missing sequences, take the most proper ATG codon as the initiation codon. The criterion for choosing the initiation codon is the same as that in step 5 of the previous section.
11. Construct a multiple alignment once again by using all sequences resulting from step 10.
12. Exclude the sequences that do not have intact N- or C-terminal portions of a gene (criterion 9 in Fig. 1). For a C-missing sequence, discard it if there are five or more gaps in the first TM region in the alignment. For an N-missing sequence, discard it if there are five or more gaps in the seventh TM region. All the remaining sequences after this exclusion process are regarded as truncated genes.
13. Exclude all the truncated genes from the sequences obtained in step 3. All the remaining sequences are regarded as pseudogenes.

4 Notes

1. To save computational time, it is preferable to exclude highly similar sequences from a query gene set. To identify mammalian OR genes, 85 functional OR genes in human and mouse that show a 50 % or less amino acid identity to one another were used as query sequences.
2. All known functional OR genes used are more than 270 amino acids long. Therefore, the cutoff length of 250 amino acids is sufficiently shorter than any known functional OR genes. A conservative cutoff length is used to avoid incorrect exclusion of functional genes.
3. If the number of sequences is significantly higher than 400, it is preferable to separate them into several groups each of which contains less than 400 sequences, to reduce the computational time for the construction of multiple alignments.
4. If the number of sequences is higher than 200, it is preferable to separate them into several groups each of which contains less than 200 sequences, to reduce the computational time for the construction of phylogenetic trees and the calculation of bootstrap values.
5. In most cases, the Class I and Class II gene clades are supported with relatively high bootstrap values in a phylogenetic tree. However, if the distinction between Class I and Class II clades is unclear, add several mammalian Class I and Class II OR genes that are present in the database to the dataset and construct a phylogenetic tree once again.
6. The reason for using the cutoff *E*-value of 1e–20 is as follows. First, the *E*-value for a best-hit to the genomic region encoding a non-OR gene is 1e–18 or larger. Second, all best-hit sequences with the *E*-value below 1e–20 obtained by OR gene queries are more similar to OR genes than to any known non-OR genes. Therefore, OR pseudogenes and non-OR genes are distinguishable by using the *E*-value of 1e–20.

Acknowledgments

This work was supported by grant (20770192 and 23770271) from the Ministry of Education, Culture, Sports, Science, and Technology, Japan.

References

1. Niimura Y (2009) Evolutionary dynamics of olfactory receptor genes in chordates: interaction between environments and genomic contents. Hum Genomics 4:107–118
2. Niimura Y (2012) Olfactory receptor multigene family in vertebrates: from the viewpoint of evolutionary genomics. Curr Genomics 13:103–114
3. Buck L, Axel R (1991) A novel multigene family may encode odorant receptors: a molecular basis for odor recognition. Cell 65:175–187
4. Fredriksson R, Lagerström MC, Lundin LG et al (2003) The G-protein-coupled receptors in the human genome form five main families. Phylogenetic analysis, paralogon groups, and fingerprints. Mol Pharmacol 63:1256–1272
5. Young JM, Shykind BM, Lane RP et al (2003) Odorant receptor expressed sequence tags demonstrate olfactory expression of over 400 genes, extensive alternate splicing and unequal expression levels. Genome Biol 4:R71
6. Niimura Y, Nei M (2005) Evolutionary dynamics of olfactory receptor genes in fishes and tetrapods. Proc Natl Acad Sci USA 102:6039–6044
7. Niimura Y (2009) On the origin and evolution of vertebrate olfactory receptor genes: comparative genome analysis among 23 chordate species. Genome Biol Evol 1:34–44
8. Glusman G, Bahar A, Sharon D et al (2000) The olfactory receptor gene superfamily: data mining, classification, and nomenclature. Mamm Genome 11:1016–1023
9. Nei M, Niimura Y, Nozawa M (2008) The evolution of animal chemosensory receptor gene repertoires: roles of chance and necessity. Nat Rev Genet 9:951–963
10. Niimura Y, Nei M (2003) Evolution of olfactory receptor genes in the human genome. Proc Natl Acad Sci USA 100:12235–12240
11. Niimura Y, Nei M (2005) Comparative evolutionary analysis of olfactory receptor gene clusters between humans and mice. Gene 346:13–21
12. Niimura Y, Nei M (2005) Evolutionary changes of the number of olfactory receptor genes in the human and mouse lineages. Gene 346:23–28
13. Niimura Y, Nei M (2007) Extensive gains and losses of olfactory receptor genes in mammalian evolution. PLoS One 2:e708
14. Go Y, Niimura Y (2008) Similar numbers but different repertoires of olfactory receptor genes in humans and chimpanzees. Mol Biol Evol 25:1897–1907
15. Matsui A, Go Y, Niimura Y (2010) Degeneration of olfactory receptor gene repertories in primates: no direct link to full trichromatic vision. Mol Biol Evol 27:1192–1200
16. Crasto C, Marenco L, Miller PL et al (2002) Olfactory receptor database: a metadata-driven automated population from sources of gene and protein sequences. Nucleic Acids Res 30:354–360
17. Altschul SF, Madden TL, Schaffer AA et al (1997) Gapped BLAST and PSI-BLAST: a new generation of protein database search programs. Nucleic Acids Res 25:3389–3402
18. Katoh K, Kuma K, Toh H et al (2005) MAFFT version 5: improvement in accuracy of multiple sequence alignment. Nucleic Acids Res 33:511–518
19. Man O, Gilad Y, Lancet D (2004) Prediction of the odorant binding site of olfactory receptor proteins by human-mouse comparisons. Protein Sci 13:240–254
20. Saitou N, Nei M (1987) The neighbor-joining method: a new method for reconstructing phylogenetic trees. Mol Biol Evol 4:406–425
21. Takezaki N, Rzhetsky A, Nei M (1995) Phylogenetic test of molecular clock and linearized trees. Mol Biol Evol 12:823–833
22. Crooks GE, Hon G, Chandonia JM et al (2004) WebLogo: a sequence logo generator. Genome Res 14:1188–1190

Part II

Theoretical Assessments of Olfactory Receptor Structure and Function

Chapter 4

Molecular Modelling of Odorant/Olfactory Receptor Complexes

Landry Charlier, Jérémie Topin, Claire A. de March, Peter C. Lai, Chiquito J. Crasto, and Jerome Golebiowski

Abstract

Providing a rationale that associates a chemical structure of an odorant to its induced perception has been sought for a long time. To achieve this, a detailed atomic structure of both the odorant and the olfactory receptor must be known. State-of-the-art techniques to model the 3D structure of an olfactory receptor in complex with various odorants are presented here. These range from sequence alignment with known structures to molecular dynamics simulations in a realistic environment.

Key words Molecular modelling, Homology, 3D-structure, Docking, GPCR, Olfactory receptor, MM-GBSA

1 Introduction

The connection between the structure of an odorant and its role in the perception of odors has been long sought. At the molecular level, the perception of smell is rooted in the activation of olfactory neurons, each of them housing an olfactory receptor (see (1) for a review). The large number of olfactory receptors (ORs) gives rise to the idea of olfaction being associated with a combinatorial signal (2) which is, for now, virtually impossible to mechanistically elucidate in terms of the odorant's chemical structure. Prior to the discovery of olfactory receptors, Structure–Property relationship studies of odors projected promising results (3). Unfortunately, they suffered from a major limitation since the role of the ORs was not considered.

With the recent advances in both computing power and bioinformatics methodologies, molecular modelling has evolved as a force that allows us a glimpse into the nature of OR–odor interactions at a molecular level. Although one of the major quests of molecular modelling, *i.e.*, the description of a protein structure on the basis

Chiquito J. Crasto (ed.), *Olfactory Receptors: Methods and Protocols*, Methods in Molecular Biology, vol. 1003,
DOI 10.1007/978-1-62703-377-0_4, © Springer Science+Business Media, LLC 2013

of its amino acid sequence is within reach (4), ORs structural description represents a particularly challenging task, since these proteins belong to the family of G-protein-Coupled-Receptors (GPCRs)-membrane proteins, whose structures are notoriously difficult to structurally characterize.

Two distinct approaches, *ab initio* or homology modelling methods, can be used to overcome the lack of X-ray structures of ORs. *Ab initio* protocols use classical laws of physics to construct 3D structures on the basis of the amino acid sequence (see (5) for an example on GPCRs). Homology modeling methods aim at predicting the structure of a protein of interest from a set of experimentally known structures. The latter is the major methodology used to build theoretical 3D models. Here, we describe the materials and methods to build a full, atomistic 3D structure of an OR, both free or complexed with an odorant. Additionally, embedding the OR or its complex with an odorant within a solvated phospholipid bilayer and system relaxation by molecular dynamics simulations is presented. The present protocol has notably been used in ref. 6.

2 Materials

2.1 Sequence Comparison and Alignment

1. The protein sequences can be downloaded on servers like the Protein Information Resource (PIR: http://pir.georgetown.edu/) or the Human Olfactory Receptor Data Explorer (HORDE: http://genome.weizmann.ac.il/horde/) (see Chapter 2).
2. The alignment can be carried out by freeware such as Jalview (http://www.jalview.org/) or directly on servers like PIR.

2.2 3D Structure Building

1. Modeller (a homology or comparative modelling of protein three-dimensional structures freeware).
2. Any 3D-visualization software (VMD, Chimera, etc.).

2.3 Ligand Docking

1. Any docking software (AUTODOCK VINA, GOLD, etc.).
2. Files containing the 3D structure of the odorants (in *pdb* or *mol2* file format).

2.4 Membrane Embedding

1. GROMACS molecular modelling software.
2. A file containing the 3D structure of the phospholipid membrane (DPPC or POPC for example can be found on P. Tieleman's Web page: http://moose.bio.ucalgary.ca/index.php?page=Structures_and_Topologies).
3. InflateGro script to build the membrane environment around the OR (http://moose.bio.ucalgary.ca/files/inflategro).
4. Alternatively, the Maestro from Schödinger, inc. (http://schrodinger.com). The software has a membrane building protocol.

2.5 Molecular Dynamics

1. Any molecular dynamics software (AMBER, GROMACS, NAMD, etc.).
2. Choose a force field developed for both proteins and lipids.

3 Methods

3.1 Sequence Alignment

The alignment should be performed with at least one known experimental structure of a class A GPCR. Currently, the X-ray structures of at least eight height different GPCRs have been solved: the bovine rhodopsin (7), the human adenosine A2a receptor (8), the turkey beta-1 adrenergic receptor (9), the human beta-2 adrenergic receptor (10), the human CXCR4 chemokine receptor (11), the human dopamine D3 receptor (12), the human histamine receptor H1 (13), and the S1P1 sphingosine 1-phosphate receptor (14). The sequence identity between these proteins and the ORs is less than 20 % and thus, should ideally be compensated by the use of experimental data. In spite of this low sequence identity, several regions are nonetheless conserved in GPCRs, such as the *GN* residues in helix 1, the *DRY* segment in helix 3. A cysteine residue present in helix 3 and another one in the extracellular loop 2 (EL2) form a conserved cysteine bridge, observed in experimental structures. A second cysteine pair is conserved in 98 % of ORs sequences. Here it involves the cysteine residues 169 and 189. Moreover, other residues are generally conserved within the OR family (15):

- *LHXPMYFFL* in the beginning of helix 2.
- *MAYDRYVAICXPLXY* in the end of helix 3.
- *SY* in helix 5.
- *KAFSTCXSH* in helix 6.
- *PMLNPFIYSLRN* in helix 7.

1. Paste both the OR and experimentally known GPCR sequences in FASTA format in the PIR Web site (http://pir.georgetown.edu/pirwww/search/multialn.shtml). Several sequences are particularly useful for ORs studies:
   (a) The multiple alignment published by Man et al. (16) aligned five ORs with five others GPCRs (including the rhodopsin sequence). It helps to correctly align conserved regions.
   (b) The sequences of the other experimental GPCR structures (notably the human β1 and β2 adrenergic receptors).
   (c) Other studies have been performed on ORs and can complete the set (6, 17–25) (see Note 1).

2. Retrieve the alignment and open it in the Jalview software. Eventually, manually assess the alignment to ensure that the conserved regions discussed above are correctly aligned. Figure 1 shows an alignment performed to produce a structure of the human OR, *OR1G1* (also named hOR17-209).

3.2 From Sequence to 3D Structure

1. On the Protein Data Bank Web site, obtain the 3D structure of: the bovine Rhodospin (PDB id: 1U19), the human β2 adrenergic receptor (PDB id: 2RH1), the turkey β1 adrenergic receptor (PDB id: 2VT4), and the human A2a adenosine receptor (PDB id: 2YDV). Modify each PDB file by removing

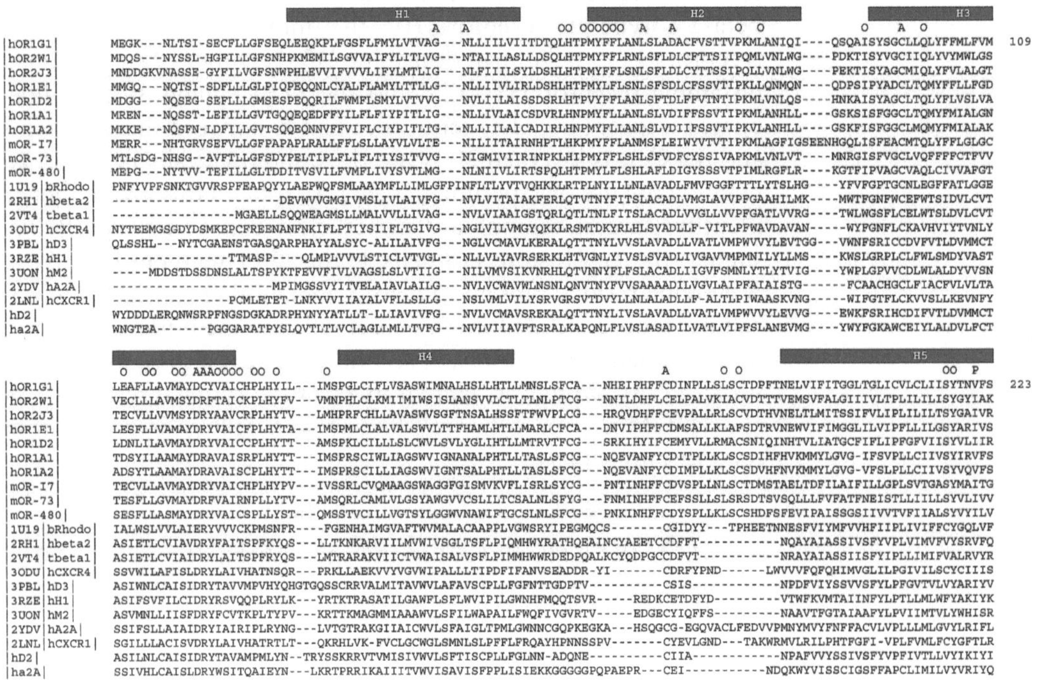

Fig. 1 Alignment of OR1G1, several ORs and other class A GPCR sequences. The predicted transmembrane helices are shown as *gray bars*. For a better reading, the first residues of mOR-480, as well as the central part of the third intracellular loop of hD2 (human dopamine D2 receptor), ha2A (human 2A adrenergic receptor) were omitted. Some residues are identified as: *O* amino acid residues common to OR, *A* amino acid residues common to class A GPCR, *G* amino acid residues allowing the GPCR activation, *C* amino acid residues including in a potential OR conformational change, *P* OR amino acid residue in contact with the G protein

water molecules, β-factor, etc., keeping only the receptor's residues and the natural ligand.

2. Prepare the MODELLER input file. Specify the cysteine residues forming S-S bridges. Note that two sulfur bridges are highly probable in ORs based on the alignment with known class A GPCRs such as β2-adrenergic receptor. In our case these two bridges involve cysteine residues 97 and 179 and also 169 and 189. Be sure to consider that the ligand (retinal in 1U19, cyanopindolol in 2VT4, carazolol in 2RH1, or ZM241385 in 3EML) located within the reference protein is inserted in the OR during the building process. It will build the OR structure with a quite large internal cavity, in which the odorants will later be docked.
3. Generate a large number of putative structures (say 50–100) to allow maximum flexibility among these structures. The template structure considered for the building procedure can be either one of the solved X-ray structures of GPCR (rhodopsin, β1, β2, or adenosine receptors, CXCR4 chemokine receptor, dopamine D3 receptor, histamine receptor, lipid GPCR, muscarinic M2 receptor, nociceptin receptor, delta opioid receptor, kappa opioid receptor, mu opioid receptor), or one obtained by using all as templates for the OR target. Typically, currently, most OR models have been built with rhodopsin as a template. One of the major differences between the crystallized GPCRs lies in the conformation of the extracellular loop 2 (EL2). This EL2 was shown to be important for ligand recognition in class A GPCRs (26) and its structure has to be considered with care. The several families of structures will be analyzed in the next steps (see Note 2).

3.3 3D Model Analysis and Validation

The analysis and validation procedure depends on both classical physicochemical considerations (hydrophobicity, hydrophilicity, etc.) and on more elaborated protocols, such as the Ramachandran plots, which represent the amino acid conformations.

1. View each model using the visualization software.
2. Eliminate the apparently badly folded structures (e.g., those with entwined loops).
3. Eliminate models where too many hydrophobic residues are in the extracellular loops.
4. Check the residues which are in a sphere of five Angstrom units around the ligand of the reference protein and compare with already studied ORs (6, 17–25). Eliminate models for which either none or too few correspondences are found with ORs built with experimental constraints (site-directed mutagenesis) (see Note 3).
5. Check the Ramachandran plot for each remaining model (see Fig. 2). They can be built on Web servers like Rampage

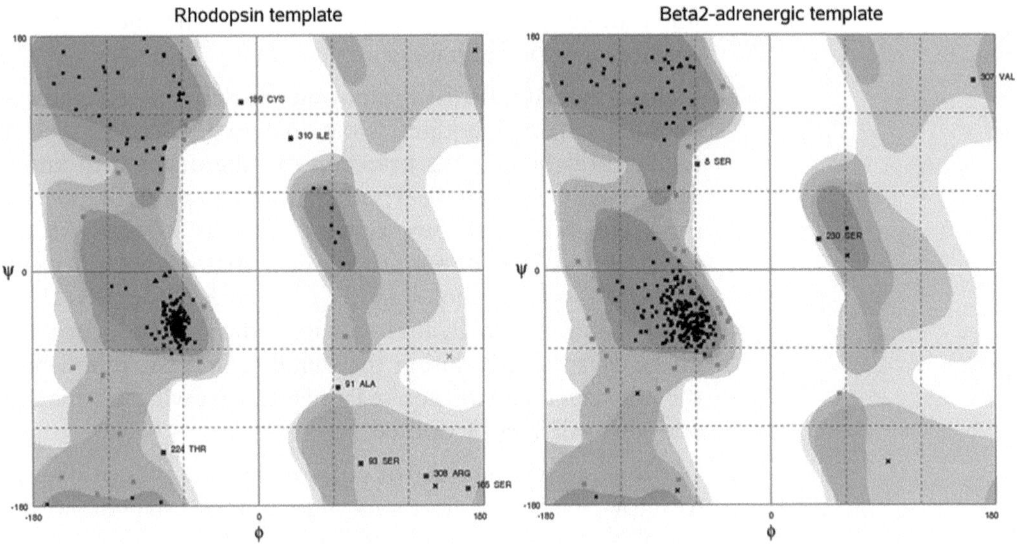

Fig. 2 Ramachandran plot for two homology models based on rhodopsin or β2-adrenergic templates

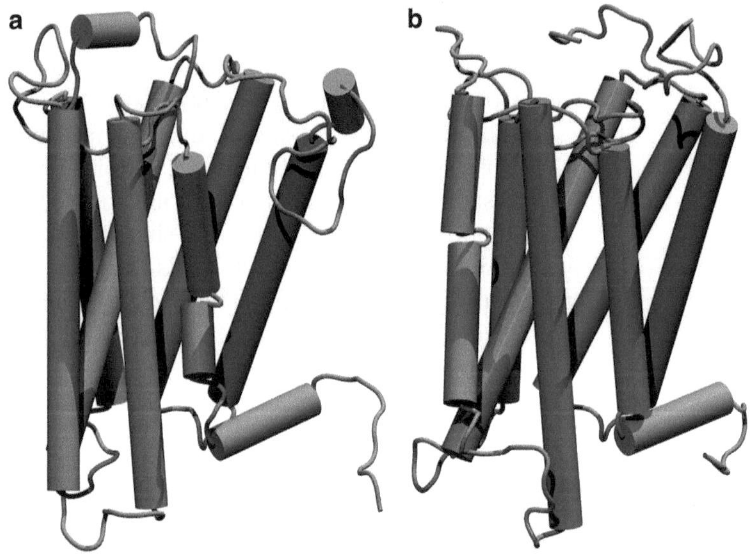

Fig. 3 OR1G1 structures built from the β2-adrenergic (PDB id: 2RH1, **a**) and rhodopsin (PDB id: 1U19, **b**) templates

(http://mordred.bioc.cam.ac.uk/~rapper/rampage.php) or on Visual Molecular Dynamics. Select models with the least residues in the outlier regions.

6. One or two structures should be chosen for each family, as shown in Fig. 3.

Figure 3 highlights small deviations between the models built from either the β2-adrenergic receptor or the rhodopsin structures. These differences arise from various kinks or shifts in the secondary

Fig. 4 Close-up view of the binding site of OR1G1 bound to camphor

structure of some helices. Moreover, variations in the sequence of the extracellular domain among the GPCRs contribute to the diversity of secondary structures. In this case, however, the residues that constitute the binding cavity are similar, rationalizing the use of these models for ligand–receptor interaction analysis purposes.

3.4 Building the Complexes

1. At this step, the ligand of the template protein is always present in the binding site. Remove it manually to enable the docking of the models of odorant molecules.
2. Dock the ligand into the binding site, either with a docking protocol or manually if sufficient evidence exists of a preferred docking configuration (see Note 4).
3. Check the environment of the odorant by comparing the ligand/receptor interactions with those described in studies involving site-directed mutagenesis or other pure *in silico* studies. Select the docking conformations involving residues found to be important in site-directed mutagenesis experiments.
4. This stage can be considered the final step, resulting in a 3D-structure of an OR bound to an odorant. Much information is already present in this model, such as the binding affinity estimation, the nature of residues forming the main contacts with the odorants and the orientation of the odorant within the binding site, as shown in Fig. 4. We describe nonetheless, additional and more technical procedures, to refine these data. The relaxation of the structure is likely to slightly change conclusions drawn from this first model. Indeed, the docking protocol has proposed several positions of the ligand within the binding site. Generally, the scoring functions lead to very weak energy differences between the different poses. This suggests that at least two or three conformations may be considered for further

refinement. A rescoring function, based on statistics accrued during Molecular Dynamics (MD) simulations should help to get a more accurate model.

3.5 Membrane Embedding

To relax the system in a realistic environment, the phospholipidic membrane, as well as the intra and extracellular medium should be modelled. These steps are rather technical since they require a good knowledge of the use of molecular simulation software.

No odorant should be present in the binding site, since the force-field does not necessary recognize the odorant atoms. The whole embedding procedure can be done with the unbound OR. The odorant can be reintroduced after the OR is stabilized in the membrane (see Note 5). Steps 1 to 7 describe a complex protocol using GROMACS. Step 8 is a much simpler alternative using Maestro.

1. Check the width of the initial membrane you have built or downloaded. Eventually duplicate it with the *genconf* command of GROMACS in directions specified by the option *-nbox*.
2. Open both the membrane and the OR files using visualization software. Check the position of the membrane with respect to the residues belonging to the OR (hydrophobic residues in the membrane, etc.). A tryptophan residue is for example a good indicator of the membrane position. Its pyrrole functional group can form hydrogen bonds with the polar heads of the lipids while its aromatic cycle interacts with the hydrophobic part of the lipids. If the membrane is badly positioned use the *editconf* command of Gromacs with the option *-center* specifying the coordinates of the center of the membrane. Use the option *-rotate* to rotate the protein around *x*, *y* or *z* and check that the OR principal axis of inertia is orthogonal to the water/lipid interface.
3. Paste the coordinates of the membrane PDB file in the PDB file of the rotated OR. At this stage, you have probably created a structure where many phospholipids have steric clashes with the receptor. Further refinement is warranted.
4. Generate the topology (*top*) and the coordinates (*gro*) files of the system OR/membrane. Use the *pdb2gmx* command of Gromacs with the *-ignh* command (ignore hydrogen). Choose the force field (for example "*Gromos 54a7*"). During the process, the cysteine bridges are recognized and created. Nonetheless, check that the appropriate cysteine residues are taken into account (if this is not the case, it indicates that the alignment prior to modelling is incorrect).
5. Inflate the lipids to eliminate steric clashes. Use the *inflategro* script. This script requires a scaling factor and a cutoff, to scale the phospholipids coordinates and deletes those that are too close to the receptor.

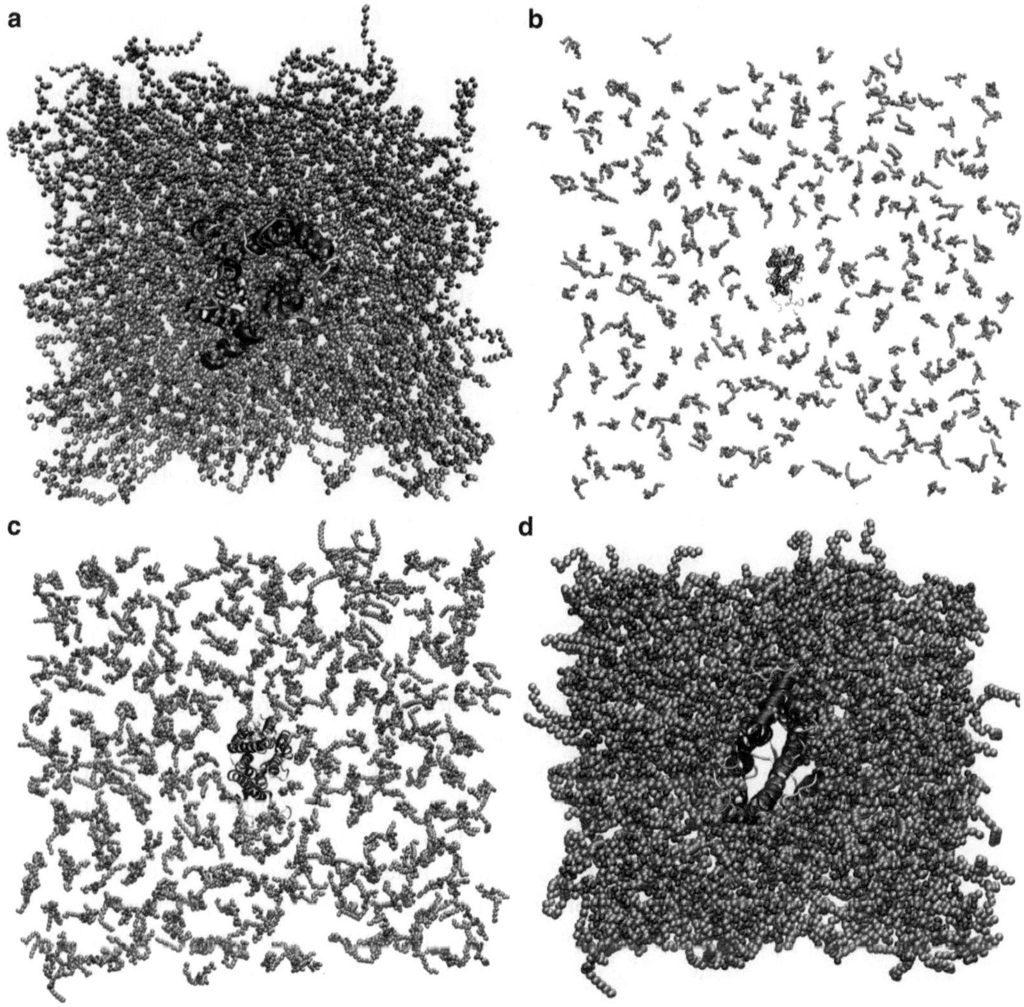

Fig. 5 Inflategro method. (**a**) Alignment of the OR on the membrane. (**b**) Inflation of the lipid. (**c**) Tenth step of deflation. (**d**) Final model

6. Slowly deflate the lipids.
   - (a) Create a minimization input file restraining the coordinates of the protein and the phosphor atom of the lipids in the three dimensions and in the *z*-dimension respectively.
   - (b) Begin minimizing (with Gromacs command *grompp* and *mdrun*).
   - (c) Use the *inflategro* script.
   - (d) Repeat the above steps until the area per lipid containing in the areaperlipid.dat file reaches the experimental values (see Note 6) (27). Figure 5 illustrates the different steps of the *inflategro* protocol.

7. Solvate your system with water molecules and add ions to neutralize the system using a default protocol.
8. Maestro (Schrödinger inc.) has a membrane building protocol that allows inserting your OR in a box containing the membrane, the water phase and the ions. It is very user-friendly but requires longer equilibration time by means of molecular dynamics.

3.6 Molecular Dynamic Simulations

The relaxation of the system implies that its energy is minimized and then heated to physiological temperature, subsequently allowing atoms and molecules to move in their environment.

1. Minimize the energy of the system in two steps. First, minimize the energy of the solvent by freezing the OR and the lipids. Second, minimize the energy of the whole system.
2. Heat the system gradually up to the desired temperature (generally 310 K), keeping the volume of the simulation box fixed.
3. Equilibrate the system at one atm. with a semi-isotropic pressure scaling.
4. The molecular dynamics simulation production phase is continued from the last equilibration step, to collect sampling for further analysis. For most odorants, the cavity created during the homology modelling is too large and tightens during the equilibration steps. The molecular dynamics production phase is then performed to evaluate the residues in contact with the odorant from a statistical point of view. These residues can be different from those found in the starting structure obtained with the docking procedure. Figure 6 illustrates the structure of hOR1G1 just after the docking step and after a 20 ns molecular dynamics simulation. Additionally, binding free energy estimation protocols can be considered to compare the computed affinity with experimental data, such as calcium imaging or dissociation constant measurements.

3.7 Docking Rescoring

Since the interaction between an odorant and an olfactory receptor are mainly hydrophobic, one can observe a reorganization of the ligand in the binding cavity during MD simulations.

To decipher the binding mode of a ligand, multiple molecular dynamics protocols can be used. According to available computational resources, several poses obtained from the docking protocol are chosen as starting points for MD simulations. Each trajectory is then analyzed using for example a MM-GBSA protocol. It has the advantage of providing a decomposition of the free energy of binding on a per-residue basis.

This method enhances the sampling of the binding cavity by the ligand, and thus allows finding the main amino acids involved

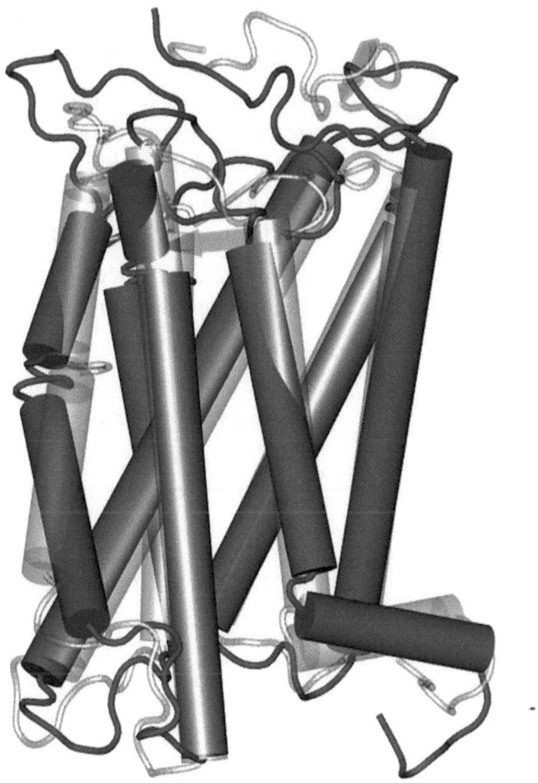

Fig. 6 Difference between the initial structure (*transparent*), obtained after the homology building procedure and the structure relaxed after 20 ns of simulation (in *dark gray*)

in the interaction. It appears to be particularly efficient in the case of nondirectional interaction between the ligand and the cavity.

For a detailed protocol of using MM-GBSA with AMBER, the reader should refer to http://ambermd.org/tutorials/advanced/tutorial3/.

4 Notes

1. Even if no OR structure is known, it is important to put several other ORs in the alignment, to be sure that the crucial conserved sequences or residues are correctly accounted for during the alignment process.
2. At this stage, one can eventually perform a geometry optimization of the residues side-chains (and only the side-chains) with molecular modelling software. Indeed, it is difficult to estimate side-chain orientations.
3. If many models fulfil only a part of the criteria (discussed in Subheadings 3.3, steps 3 and 4), this can point to a bad alignment

of the sequences. This may necessitate a modification of the alignment (Subheading 3.1).

4. You can use a standard docking protocol for this step (not described here). Generally, the olfactory receptor is considered rigid and the only the odorants can undergo conformational changes. One can consider that the binding site will be identical to those found in rhodopsin and adrenergic receptors.
5. Membrane embedding can be done by several methods. Reference 28 summarizes them.
6. It is very useful to use a script at this stage. Indeed, the deflation of the system takes lot of simulation steps (depending on your scaling factor during the inflation).

Acknowledgments

The CINES is acknowledged for providing computer time for the project cmi1024.

References

1. Breer H (1994) Odor recognition and second messenger signalling in olfactory receptor neurons. Semin Cell Dev Biol 5:25–32
2. Malnic B, Hirono J, Sato T, Buck L (1999) Combinatorial receptor codes for odorants. Cell 96:713–723
3. Rossiter KJ (1996) Structure-odor relationships. Chem Rev 96:3201–3240
4. Cozzetto D, Tramontano A (2008) Advances and pitfalls in protein structure prediction. Curr Protein Pept Sci 9:567–577
5. Abrol R, Bray JK, Goddard WA III (2011) BiHelix: towards de novo structure prediction of an ensemble of G-protein coupled receptor conformations. Proteins 80:505–518
6. Charlier L, Topin J, Ronin C, Kim SK, Goddard WA III, Efremov R, Golebiowski J (2012) How broadly tuned olfactory receptors equally recognize their agonists. Human OR1G1 as a test case. Cell Mol Life Sci 69:4205–4213. doi:10.1007/s00018-012-1116-0
7. Okada T, Sugihara M, Bondar A-N, Elstner M, Entel P, Buss V (2004) The retinal conformation and its environment in rhodopsin in light of a new 2.2.ANG. crystal structure. J Mol Biol 342(2):571–583
8. Jaakola V-P, Griffith MT, Hanson MA, Cherezov V, Chien EYT, Lane JR, Ijzerman AP, Stevens RC (2008) The 2.6 angstrom crystal structure of a human A2A adenosine receptor bound to an antagonist. Science 322(5905):1211–1217
9. Warne T, Serrano-Vega MJ, Baker JG, Moukhametzianov R, Edwards PC, Henderson R, Leslie AGW, Tate CG, Schertler GFX (2008) Structure of a beta 1-adrenergic G-protein-coupled receptor. Nature 454(7203): 486–491
10. Cherezov V, Rosenbaum Daniel M, Hanson Michael A, Rasmussen Soren GF, Thian Foon S, Kobilka Tong S, Choi H-J, Kuhn P, Weis William I, Kobilka Brian K, Stevens Raymond C (2007) High-resolution crystal structure of an engineered human beta2-adrenergic G protein-coupled receptor. Science 318(5854): 1258–1265
11. Wu B, Chien EY, Mol CD, Fenalti G, Liu W, Katritch V, Abagyan R, Brooun A, Wells P, Bi FC, Hamel DJ, Kuhn P, Handel TM, Cherezov V, Stevens RC (2010) Structures of the CXCR4 chemokine GPCR with small-molecule and cyclic peptide antagonists. Science 330(6007): 1066–1071
12. Chien EY, Liu W, Zhao Q, Katritch V, Han GW, Hanson MA, Shi L, Newman AH, Javitch JA, Cherezov V, Stevens RC (2010) Structure of the human dopamine D3 receptor in complex with a D2/D3 selective antagonist. Science 330(6007):1091–1095
13. Shimamura T, Shiroishi M, Weyand S, Tsujimoto H, Winter G, Katritch V, Abagyan R, Cherezov V, Liu W, Han GW, Kobayashi T, Stevens RC, Iwata S (2011) Structure of the human dopamine D3 receptor in complex with

a D2/D3 selective antagonist. Nature 475(7354):65–70

14. Hanson MA, Roth CB, Jo E, Griffith MT, Scott FL, Reinhart G, Desale H, Clemons B, Cahalan SM, Schuerer SC, Sanna MG, Han GW, Kuhn P, Rosen H, Stevens RC (2012) Crystal structure of a lipid G protein-coupled receptor. Science 335(6070):851–855
15. Zozulya S, Echeverri F, Nguyen T (2001) The human olfactory receptor repertoire. Genome Biol 2(6):18
16. Man O, Gilad Y, Lancet D (2004) Prediction of the odorant binding site of olfactory receptor proteins by human-mouse comparisons. Protein Sci 13(1):240–254
17. Schmiedeberg K, Shirokova E, Weber H-P, Schilling B, Meyerhof W, Krautwurst D (2007) Structural determinants of odorant recognition by the human olfactory receptors OR1A1 and OR1A2. J Struct Biol 159(3):400–412
18. Doszczak L, Kraft P, Weber H-P, Bertermann R, Triller A, Hatt H, Tacke R (2007) Prediction of perception: probing the hOR17-4 olfactory receptor model with silicon analogues of bourgeonal and lilial. Angew Chem Int Ed 46(18):3367–3371
19. Hall SE, Floriano WB, Vaidehi N, Goddard WA III (2004) Predicted 3-D structures for mouse I7 and rat I7 olfactory receptors and comparison of predicted odor recognition profiles with experiment. Chem Senses 29(7):595–616
20. Abaffy T, Malhotra A, Luetje CW (2007) The molecular basis for ligand specificity in a mouse olfactory receptor: a network of functionally important residues. J Biol Chem 282(2): 1216–1224
21. Hummel P, Vaidehi N, Floriano WB, Hall SE, Goddard WA III (2005) Test of the binding threshold hypothesis for olfactory receptors: explanation of the differential binding of ketones to the mouse and human orthologs of olfactory receptor 912-93. Protein Sci 14(3):703–710
22. Floriano WB, Vaidehi N, Goddard WA III, Singer MS, Shepherd GM (2000) Molecular mechanisms underlying differential odor responses of a mouse olfactory receptor. Proc Natl Acad Sci U S A 97(20):10712–10716
23. Singer MS (2000) Analysis of the molecular basis for octanal interactions in the expressed rat I7 olfactory receptor. Chem Senses 25(2): 155–165
24. Baud O, Etter S, Spreafico M, Bordoli L, Schwede T, Vogel H, Pick H (2011) The mouse eugenol odorant receptor: structural and functional plasticity of a broadly tuned odorant binding pocket. Biochemistry 50(5):843–853
25. Gelis L, Wolf S, Hatt H, Neuhaus EM, Gerwert K (2012) Prediction of a ligand-binding niche within a human olfactory receptor by combining site-directed mutagenesis with dynamic homology modeling. Angew Chem Int Ed 51:1274–1278
26. Avlani VA, Gregory KJ, Morton CJ, Parker MW, Sexton PM, Christopoulos A (2007) Critical role for the second extracellular loop in the binding of both orthosteric and allosteric G protein-coupled receptor ligands. J Biol Chem 282:25677–25686
27. Nagle JF, Tristram-Nagle S (2000) Structure of lipid bilayers. Biochim Biophys Acta 1469(3):159–195
28. Kandt C, Ash WL, Tieleman DP (2007) Setting up and running molecular dynamics simulations of membrane proteins. Methods 41(4):475–488

Chapter 5

The Electrical Properties of Olfactory Receptors in the Development of Biological Smell Sensors

Eleonora Alfinito, Jeremy Pousset, and Lino Reggiani

Abstract

We present here the results of the investigation of the electrical properties of two olfactory receptors (ORs): rat, OR I7, and human, OR 17-40, which are of interest in the creation of smell nanobiosensors. Described here is our investigation comparing the results from experiments using electrochemical impedance spectroscopy with the theoretical predictions obtained from a recently developed impedance network protein analog. The changes in the OR response following excitation correlated with the protein conformational change. The satisfactory agreement between theory and experiment points to a promising development of a new class of nanobiosensors based on the electrical properties of sensing proteins.

Key words Olfactory receptors, Biosensors, Protein electrical properties

1 Introduction

Olfactory receptors (ORs) are transmembrane proteins belonging to the family of G protein coupled receptors (GPCRs). In mammals, ORs are responsible for the first step in the process leading to olfaction. Because of their selectivity and sensitivity, which enable the organism to discriminate thousands of different odors, ORs represent the gold standard for the development and creation of a smell nanobiosensor.

Olfaction in mammals is complex and archaic. It is fundamental to rapidly identifying preys and predators and, in conjunction with taste, necessary for avoiding toxic food. Acknowledging the complexities and variability in this sense has been key in promoting efforts towards conceptualizing and developing a novel smell sensor, a notion that has received support from the bioelectronic olfactory neuron device (BOND) project (1).

Conventional smell sensors, usually called *electronic noses*, have been in production for several years (2, 3). They usually achieve detection by means of a solid-state sensitive component, translating

Chiquito J. Crasto (ed.), *Olfactory Receptors: Methods and Protocols*, Methods in Molecular Biology, vol. 1003,
DOI 10.1007/978-1-62703-377-0_5,

the captured information into an electrical signal. The signal is then digitalized and elaborated by appropriate algorithms.

Currently, the sensitivity and selectivity of these devices are high but do not come close to mimic the mammalian nose. To improve detector features, many efforts are currently devoted to create biomimetic smell sensors, which can integrate the sensitivity and specificity of ORs with leading and user-friendly electronic devices (1, 2, 4–8).

Our aim is to produce a high-selectivity sensor. Since we do not consider it practicable to reproduce the complexity of the in vivo olfaction system, we aim to specifically target a unique parameter in the process of olfaction sufficient to give the notice of capture. This parameter is the conformational change ORs undergo when they bind one of their specific odorant molecules (ligands). It has been conjectured that the conformational change is already sufficient to modify the electrical properties of a protein appropriately anchored between Ohmic contacts. The validation of this conjecture represents a proof of concept that a nanobiosensor might parallel olfaction in nature.

In the following sections, we review recent experiments on the electrical characterization of ORs and describe a theoretical and computational strategy to provide a microscopic interpretation of the results. The final considerations illustrate our perspectives on the feasibility of developing a smell sensor based on ORs.

2 Methods

Olfactory receptors belong to the large family of GPCRs (9, 10), proteins devoted to infer information from the environment (the presence of light, odors, pheromones, toxins, etc.) and to transmit it to the nervous system. Therefore, they are crucial to the health and the safety of their host organisms. For some time, this has prompted broad-spectrum investigations of OR, including: genetic analysis (11–15), ranging from identification and expression, to the development of technologies that measure OR excitation (8, 16–25).

GPCRs, and specifically ORs, while possessing varying sequences (in order to discriminate different odors), nevertheless, fold into quite similar tertiary structures, consisting of seven transmembrane helices (see Fig. 1).

While X-ray crystallography and NMR techniques have been used to deduce the tertiary structure of proteins (26), structure determinations of GPCRs are challenging and only a few GPCRs have been structurally characterized (27). By using homology modeling, however, it is possible to create a plausible structure of a GPCR, from a physicochemical point of view (28–30). Specifically, ORs are modeled by using as a template, bovine rhodopsin, a light receptor, whose structural features have been widely investigated

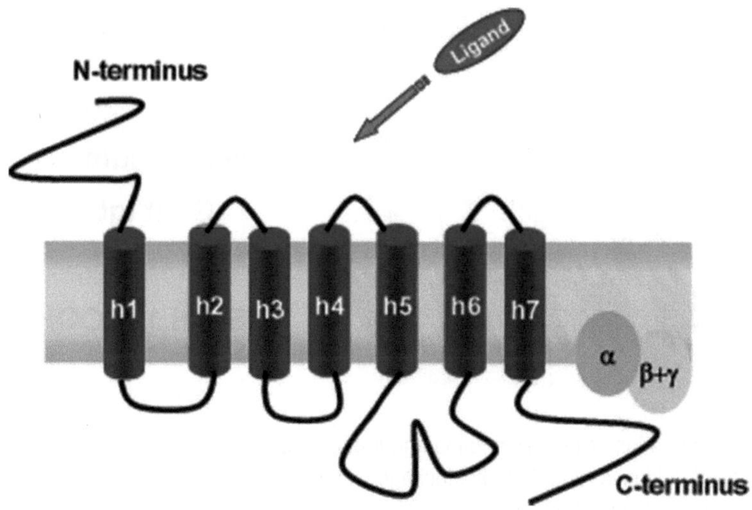

Fig. 1 Schematic drawing of a GPCR. The *rectangular box* represents the cellular membrane which separates the extra-cellular region (above) from the cytoplasmic region (below). The seven *cylinders* show the seven transmembrane α-helical domains of the receptor, while the loops and the two termini (N-terminus and C-terminus) are shown by *black lines*. The two *ellipses* represent the subunits of the G-protein and, finally, the *ellipse* on the *top* shows a ligand captured by the receptor

(10). The sequence identities between bovine rhodopsin and ORs allow one to produce a homology driven model (28–30). Furthermore, the natural abundance and availability of bovine rhodopsin, in contrast to GPCRs, which are typically expressed at a very low level in cells, make this protein particularly attractive as the test receptor. Rhodopsin can be thus used to establish biotechnological platforms for chip-based screening technologies, to characterize GPCRs function and interactions (20, 31–36).

2.1 Structure and Sensing Action

In a living cell, the detection and transduction process begins with a conformational change in the GPCR following the capture of the specific ligand. This structural change activates the G protein, resulting in a complex sequence of biological mechanisms that end with the production of a neuronal electric pulse. Different ligands may induce and stabilize distinct structural conformational states, promoting varied G protein activations and receptor desensitization/internalization. Moreover, GPCRs can interact with other proteins through their C-terminal domains. They can dimerize or form heterodimers with other membrane-bound proteins depending on the function.

Understanding these mechanisms is crucial for new drug design. The drug discovery process and the research of novel GPCR-based therapeutics however, require a large body of data to allow the identification of new receptor ligands. There is an increasing need to develop genetic and chemical high-throughput screening methods for an efficient sorting of compounds with

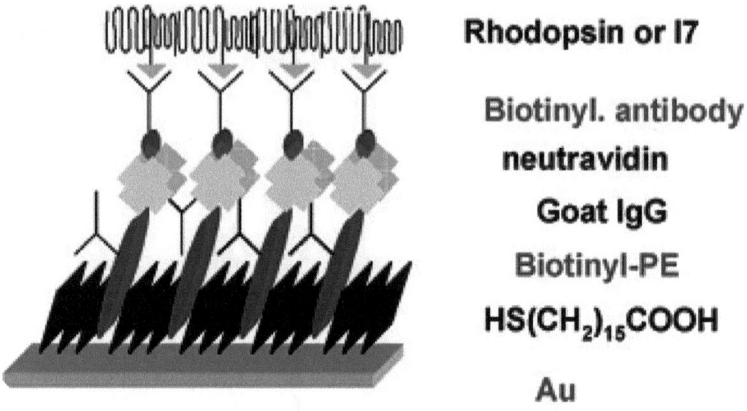

Fig. 2 Schematical illustration of the assembling and immobilization process of rhodopsin and OR I7. Step I: mixed SAMs modified gold electrode; Step II: blockage with goat IgG; Step III: binding of neutravidin; Step IV: immobilization of biotinylated antibody Biot-Rho-1D4 (17)

pharmaceutical potential. In this context, the production of GPCR-based nanobiosensors can be very useful. In nanobiosensors, the detection process bypasses the complicated sequence of biological events which follows G-protein activation. As a consequence, it is possible to achieve a sensitive and reliable monitoring of the conformational change by a direct analysis of the optical and/or the electrical response of the nanodevice itself.

2.2 Electrical Characterization

The electrical properties of living matter and, in particular, of proteins (37), are the object of increasing interest (38). Beginning with the seminal papers that described correlations between neuronal function with a network of ladder impedances (37), much work has been done on the physicochemical aspects of this correlation.

The impedance properties of some GPCRs have been investigated by using both electrochemical impedance spectroscopy (EIS) measurements and also surface plasmon resonance (SPR) (17, 18, 21, 22). These techniques, when applied to ORs, pointed to a change of impedance in the presence/absence of the specific odorant molecules. Preliminary investigations were performed on bovine rhodopsin, in order to establish an extendible protocol for such measurements (17).

We applied this procedure to two ORs: rat OR I7 and human OR 17-40, whose functional analysis—identifying odorants that bind and excite them—has been well established (18, 25, 39). EIS measurements were carried out on proteins anchored on a self-assembled multilayer (SAM), at different stages of multilayer formation (17) (see Fig. 2).

The results were described in terms of Nyquist plots (Fig. 3). Precisely, these plots can be interpreted in terms of a simple electrical analog, the Randles cell. This is the circuital description

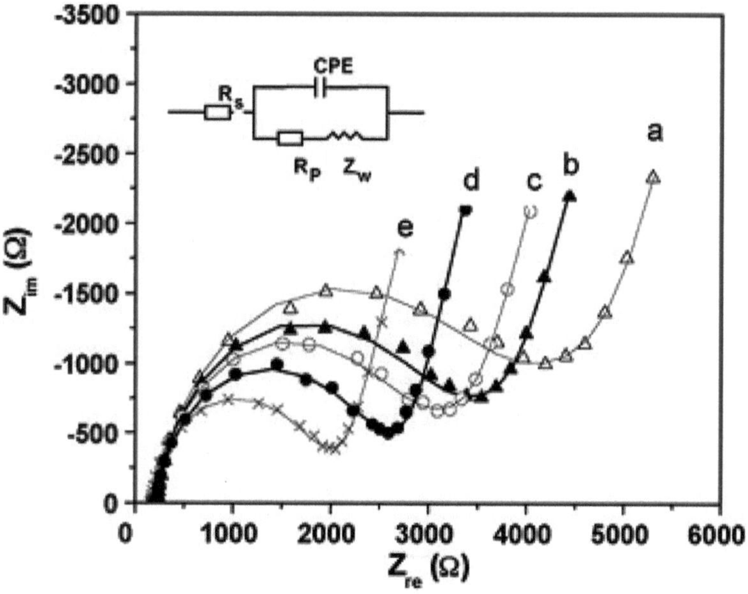

Fig. 3 Electrochemical characteristics of rhodopsin SAM at different levels of assembling with the corresponding circuital analog (Randles cell) (on the *top*). (*a*) Step I: mixed SAMs modified gold electrode; (*b*) Step II: blockage with goat IgG; (*c*) Step III: binding of neutravidin; (*d*) Step IV: immobilization of biotinylated antibody Biot-Rho-1D4; (*e*) after injection of 80 ng/ml rhodopsin membrane fraction (17, 34)

of the macroscopic properties of the electrochemical cell: The resistance of the electrolyte solution (*R*s), the polarization resistance (*R*p), and the constant phase elements (CPE) describing the electrode–electrolyte interface features, and finally the Warburg impedance (*Z*w) related to the low frequency part of the spectrum. Different levels of assembly correspond to different Nyquist plots, but only the semicircular part modifies its shape. This means that in the circuit analog, the series resistance and the Warburg impedance do not change their value significantly when compared with the polarization resistance. Indeed, both *Z*w and *R*s mainly pertain to the experimental environment. Accordingly, what really describes the sampling modification is only the impedance related to the *R*p–CPE parallel circuit.

Measurements carried out on OR I7 were related to the dose responses following interactions with heptanal and octanal, two organic compounds differing in length by one methyl group (21, 22, 39) and the sensitivity to nonbinding odorants (the helional, specific for the human OR 17-40, was tested against OR I7). In all cases, the anchored proteins show the expected behavior: An increasing activation for increasing dose of odorant, a good resolution at low concentrations (10^{-11} M), and an efficient resolution on a wide range of odorant concentration (10^{-11} M–10^{-4} M).

A similar strategy used with human OR 17-40, showed a very narrow window of detectable responses with respect to the odor concentration (19, 21). In particular, it was observed that the polarization resistance showed a peculiar bell-shaped behavior, at increasing values of the specific odorant concentration (22). The peak was centered at low concentrations of the specific ligand (around 10^{-10} M). Furthermore, by using complementary techniques, like SPR and the differential bioluminescence response, in addition to the first peak a second maximum response was observed at a higher concentration of odorant (around 10^{-5} M) (22). These results were also confirmed by differential conductance measurements (8). Such early investigations show that EIS technique is in fact useful to determine the skill of ORs to respond to the specific ligands. Furthermore, EIS also shows that the kind of response is inherently dependent on the specific OR chosen.

3 Theoretical Model

The aim of this section is to determine the variation of the protein impedance spectrum (20) subsequent to a conformational change, as shown in Fig. 4, by taking into account only geometrical changes in the tertiary structure of the protein.

We assume that the 3D structures of the protein in its native (without ligand) and activated (with ligand) states are known from the protein database (PDB) available in the literature (27) or from

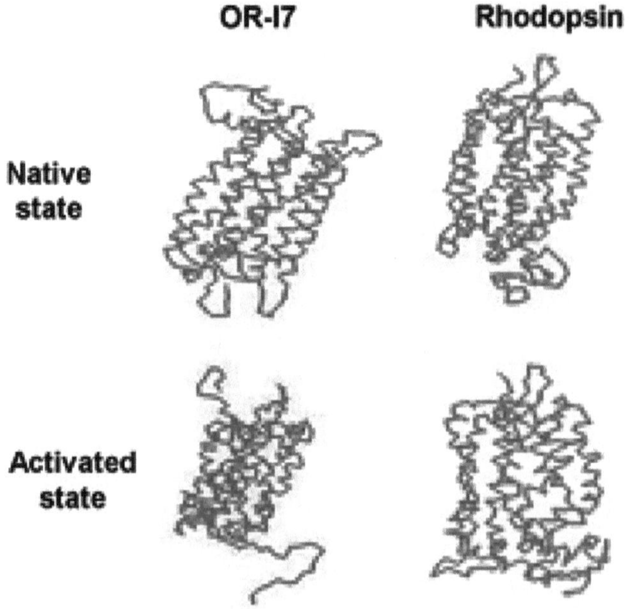

Fig. 4 Sketch of the backbone of native and activated states of bovine rhodopsin and rat OR-I7, as obtained with MODELLER software (28, 29, 34, 36)

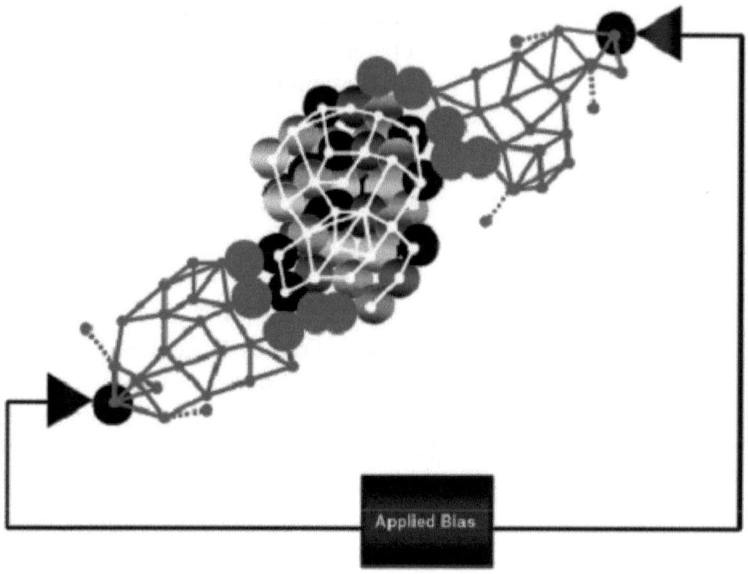

Fig. 5 Cartoon of a protein embedded into an impedance network

homology modeling (28–30). We used homology-modeled structures for the ORs described in this work (28, 29, 36).

Our procedure consists of two steps: First, the protein in an assigned state is represented by means of a topological network (graph), as sketched in Fig. 5. Each node of this graph corresponds to an amino acid (specifically to the α-carbon). If two amino acids are closer than an assigned cutoff distance, *R*c, then a link is drawn between the corresponding nodes. Accordingly, *R*c can be viewed as the radius of the *sphere of interaction* centered in each node, in such a way two nodes interact only if their spheres intersect (20, 31–36, 40–48).

The distance *R*c definitely is a measure of shape of the network, which here can be considered as an adjustable parameter of the model. Accordingly, the choice of its value has to be determined by some physical constraints and finally fixed by comparison with experiments. At this stage, we limit ourselves to a network built up with a strategy analog to that of random graphs, by using *R*c, instead of a constant probability, to select the effective links (49). Therefore, using this model we can recover some of the results obtained in the context of random and complex networks.

Second, we establish the kind of interaction represented by the links. This choice, in turn, fixes the kind of response which can be calculated. For example, by assigning a spring constant to each link, it is possible to study the network normal modes, in a way resembling the elastic network models (47, 50, 51).

In describing static and dynamic electrical responses, each link mimics a privileged channel for the charge transfer and/or charge polarization (20). In particular, it is replaced by an elementary impedance consisting of an Ohmic resistance connected in parallel with a parallel plate capacitor.

The elementary impedance between nodes *i* and *j*, *Zi,j*, is given explicitly by (20):

$$Z_{i,j} = \frac{l_{i,j}}{A_{i,j}} \frac{1}{\rho^{-1} + i\varepsilon_{i,j}\varepsilon_0\omega}.$$

where $A_{i,j} = \pi(R_c^2 - l_{i,j}^2 / 4)$, is the cross-sectional area between two spheres of radius *Rc* centered on the *i*-th and *j*-th node, respectively; $l_{i,j}$ is the distance between these centers, ρ is the resistivity, taken to be the same for every amino acid, with the indicative value of $\rho = 10^{10}$ Ωm; $i = \sqrt{-1}$ is the imaginary unit, ε_0 is the vacuum permittivity, ω is the circular frequency of the applied voltage. The relative dielectric constant of the couple of *i*-th and *j*-th amino acid, $\varepsilon_{i,j}$, is expressed in terms of the intrinsic polarizability of each amino acid (52).

This step turns the graph into an impedance network.

Finally, by taking appropriate contacts and applying an external bias, the electrical network is solved within a linear circuit analysis and its global impedance spectrum is calculated within the frequency range 1 mHz–100 kHz (20, 31–34, 36, 40, 42, 43).

For a given interaction radius, the spectrum depends on the network structure. The change of the protein state, by implying a change in the network structure, leads to a different spectrum of the global impedance. Such a variation of the impedance spectrum is here taken as an estimate of the electrical response consequent to the protein conformational change. Since the resistivity associated with an amino acid is taken, in a first approximation, to be independent of geometry and applied voltage, the above estimate should be considered as a minimum threshold value.

By considering more detailed microscopic mechanisms of charge transport (53, 54), like tunneling, and/or the possible influence of the medium surrounding the protein, the electrical sensitivity to the conformational change can be further modulated and, in particular, amplified.

The role of the medium (the lipid layer) surrounding the protein is here neglected. Indeed, the medium does not affect the OR conformational change and its effect is canceled when considering the impedance variation.

4 Results

In this section we apply the theoretical model to the investigation of the protein topology associated with a given value of the interaction radius and of the small-signal impedance associated with the change of conformation.

4.1 Topological Investigation

Here we report results obtained by using contact maps of the network. These are 2D maps of the network of connected parts, and can be drawn by assigning to each couple of linked amino acids, say *i,j*, a point of coordinates *i,j*. Two amino acids are linked only if the distance between them is less than *Rc*, the interaction radius. Therefore, by comparing the contact maps of the protein in the native and activated state we obtain a direct view of the main changes in the protein topology induced by the sensing action.

Figures 6 and 7 show the contact maps for OR I7 determined at two *Rc* values, 6 Å and 12 Å, respectively. From these figures we observe that by increasing the value of *Rc*, the density of points and thus the connectivity increase substantially, as expected. In particular, the main differences between the native and activated states are in the increased connectivities among the helices *h3-h4-h6*, where the principal binding pocket is presumably located (36, 55).

Notably, significant differences between native and activated states survive for large *Rc*, say up to *Rc* values of about 50 Å (45). A quite similar behavior has been observed for human OR 17-40 (36).

4.2 Resistance and Impedance Spectrum

Figure 8 reports the flowchart for the numerical calculation of the small-signal impedance spectrum. In brief, the steps are summarized as follows.

A. The input data are constituted by the 3D structure of the protein, in the native and activated state, the cutoff distance, *Rc*,

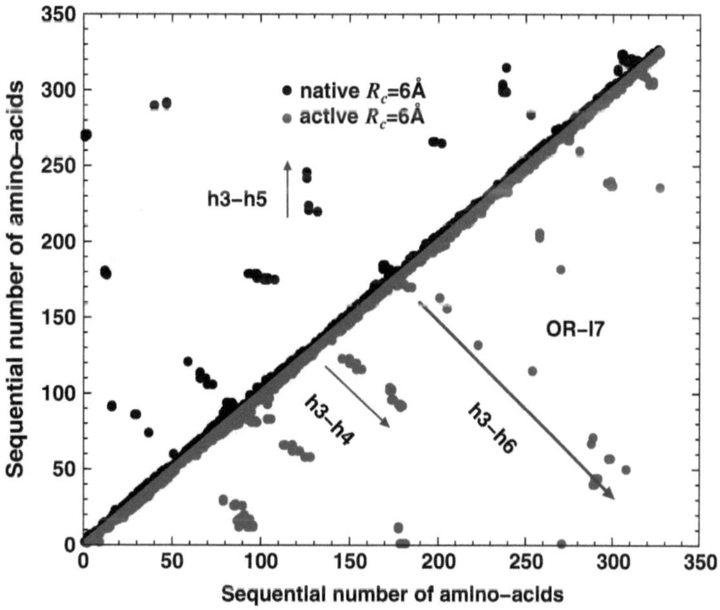

Fig. 6 Contact map of the two OR I7 representations, native state (*circles* on the *left*), and activated state (*circles* on the *right*). Because of the axial symmetry with respect to the diagonal exhibited by data, the symbols pertaining to the native map are reported only on the *left hand side* of the diagonal and those pertaining to the activated map on the *right hand side*. The interaction radius is $R_c = 6$ Å

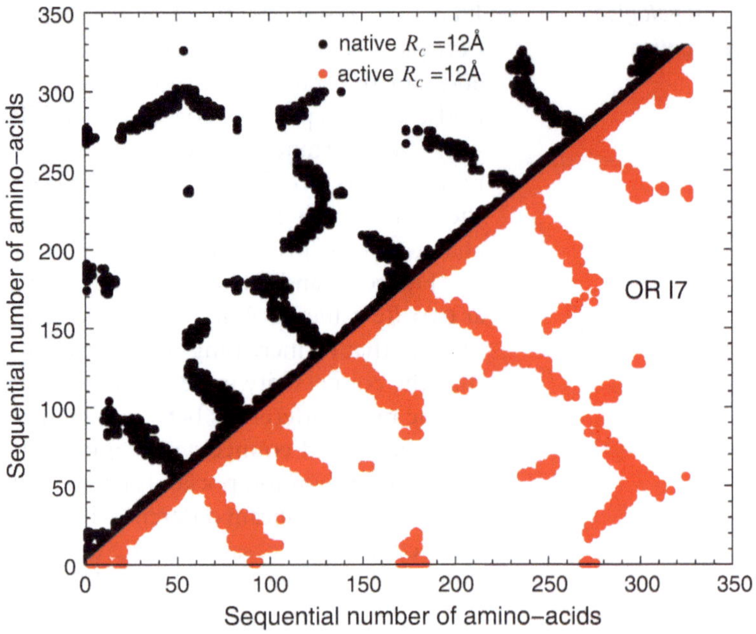

Fig. 7 The same as in Fig. 6 for an interaction radius of $R_c = 12$ Å

the frequency range and the bias value. The 3D structures are deduced by the X-ray/NMR/MODELLER files.

B. The network is created by the comparison of the amino acid distances with the cutoff value, *R*c.

C. The local impedance is calculated for each of the network links.

D. The total network impedance is calculated within the standard Gauss–Jordan method (33).

E. The process is iterated for different frequency values.

The differences between native and activated states of the considered ORs are evidenced in the change of the receptor global resistance, *R*. Figure 9 shows the shape of the relative variation of resistance, *R*, as a function of *R*c, for both proteins. The main result reported by this figure is the larger sensitivity of the rat OR I7 when compared to that of the human OR 17-40. In the former case it is possible to resolve a maximum difference of about 60% while in the latter case the maximal resolution is about 20%.

The region of maximal sensitivity is the same for both proteins, corresponding to *R*c in the range 6–14 Å.

In the case of OR 17-40, the activated state shows values of resistance which are significantly lower than those of OR I7. Furthermore, for an *R*c above about 18 Å, calculations evidence for OR 17-40 an inversion of the resistance variation, with the activated state becoming less resistive than the native state in two regions of

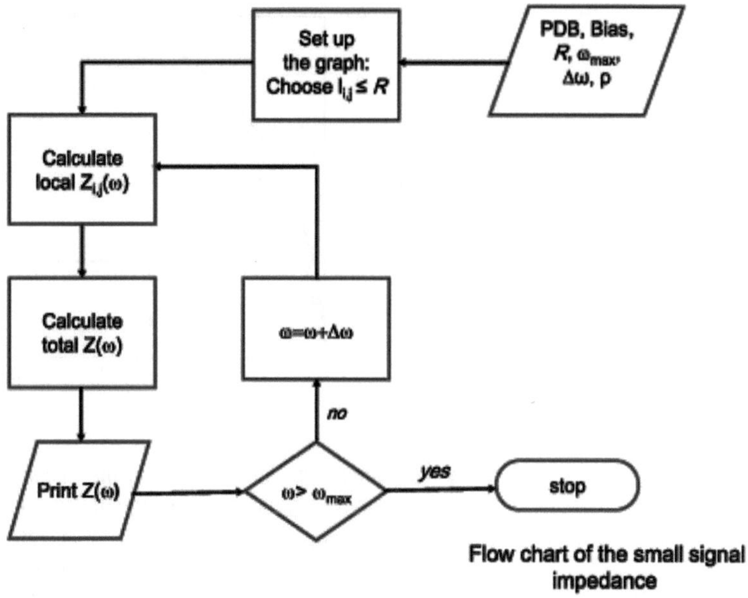

Fig. 8 Flow chart for the numerical evaluation of the small-signal impedance of a single protein

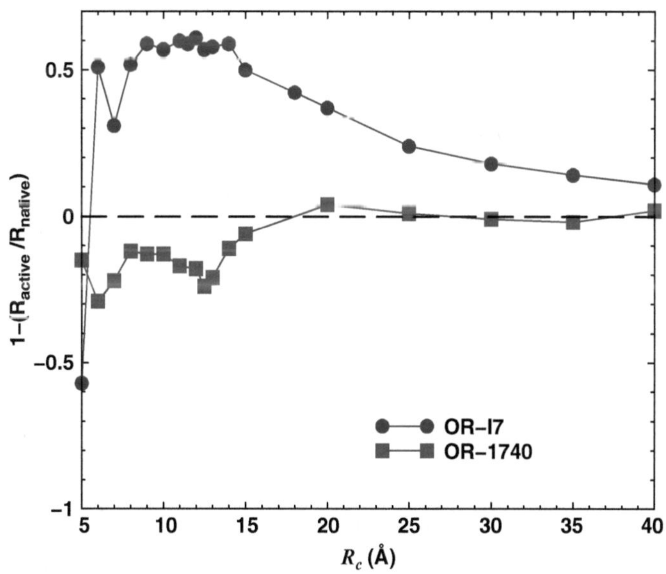

Fig. 9 Relative resistance variation for increasing *Rc* values. *Circles* refer to rat OR I7, *squares* refer to human OR 17-40

Rc values 18–24 Å and 40–60 Å, respectively. In terms of the electrical network, such an inversion is interpreted as a stronger increase of parallel connections when compared to series connections. Accordingly, the different behavior shown by the two receptors illustrates differences in the peculiarities of their network structure.

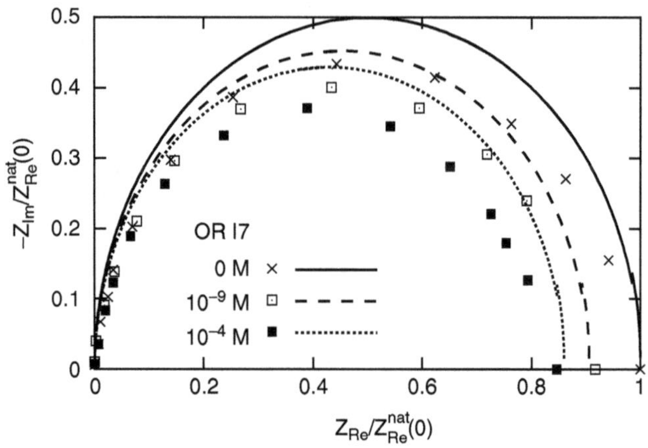

Fig. 10 Nyquist plot for rat OR I7 in the absence and presence of one of the specific ligands, octanal (45). The impedances are normalized to the static value of the native state, $Z^{nat}_{Re}(0)$ (17, 21). Symbols refer to experiments with: *Crosses* refer to no odorant, *empty squares* to an odorant concentration of 10^{-9} M, and *full squares* to an odorant concentration of 10^{-4} M. Lines refer to the theoretical results with: *Continuous curve* refers to the native state configuration, *Rc* = 50 Å; *long dashed line* refers to the activated state configuration, *Rc* = 30 Å; *long dashed line* refers to the activated state configuration, *Rc* = 25 Å

The impedance response of single proteins is explored for a wide range of frequencies, and the results are given by means of the Nyquist plot. This plot is obtained by graphing the negative imaginary part versus the real part of the global impedance, within the considered experimental (17, 18, 21).

Figure 10 illustrates the Nyquist plots with the impedance normalized to the static value of its native state for rat OR I7. Symbols refer to EIS measurements (18, 44, 45), where the solution resistance is subtracted. Curves refer to theoretical results where the single protein is representative of the entire sample. Crosses pertain to experiments carried out in the absence of the specific odorant, octanal; empty and full squares pertain to the different concentrations of the odorant as reported in the figure.

For our model, the Nyquist plots are obtained by using as input data the networks corresponding to the native and activated states at the cutoff radius that fits the experimental change of $Z_{Re} / Z^{nat}_{Re}(0)$ at the given odorant concentrations (Fig. 9).

We conjecture that the existence of a correlation between the concentration of odorant and *Rc* is due to the modifications undergone by the protein ensemble in the presence of a given concentration of odorant (56–58). The agreement between theory and experiments is found to be satisfactory from both a qualitative and quantitative point of view. The near ideal semicircle shape of the experimental Nyquist plot is well reproduced. It confirms that the

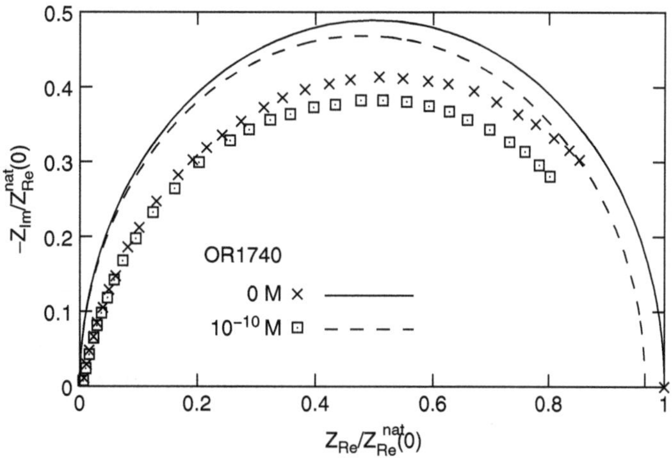

Fig. 11 Nyquist plot of human OR 17-40 in the absence and in the presence of the specific ligand (helional). The impedances are normalized to the static value of the native state, $Z_{Re}^{nat}(0)$ (45). Symbols pertain to experiments with: *Crosses* referring to no odorant, *empty squares* to an helional concentration of 10^{-10} M at room temperature (17, 21, 22). Curves pertain to theoretical results with: *Continuous curve* referring to the native state configuration as input data with *Rc* = 70 Å; *dashed line* referring to the activate state configuration with *Rc* = 46 Å

network impedance model behaves closely to a single *RC* parallel circuit. This is as expected because of the presence of a rather uniform distribution of time constants associated with the different values of the resistance and capacitance of the links (33).

Figure 11 shows analogous results for the case of human OR 17-40, in the native state and in the activated state with a concentration of 10^{-10} M of the specific odorant helional (21, 22).

Again, the agreement between theory and experiments is found to be satisfactory. The scarcity of experimental data and the lack of a certified knowledge of the 3D structures of the considered proteins lead us to consider these results as a first but significant step towards a microscopic modeling of the electrical properties of olfactory receptors.

4.3 Current–Voltage Characteristics

Current–voltage (*I–V*) characteristics of a single protein are calculated by implementing the numerical code of the small-signal impedance and introducing sequential tunneling between different amino acids as the microscopic mechanism of charge transfer (35, 41, 46, 59). The theory is validated by recent experiments carried out on mono-layered structures of bacteriorhodopsin (bR), a light-activated protein, where there is evidence of a substantial increase of the current, at a given voltage, when passing from dark to green light (35, 41, 46, 59).

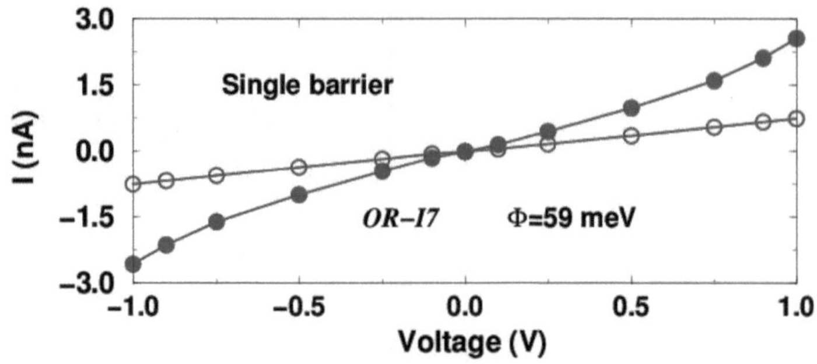

Fig. 12 *I–V* characteristics of the native and activated states of rat OR-I7. *Empty circles* refer to the native state, and *full circles* to the activated state, respectively. The *continuous lines* are guides for eyes. Data are calculated with a single barrier height $\Phi = 59$ meV

In Fig. 12, the *I–V* characteristics of OR I7 are found to predict a substantial super-Ohmic behavior at increasing applied voltages better evidenced for the activated than the native state of the protein. The parameters used for this simulation are the same working for bacteriorhodopsin: $Rc = 6$ Å and an average barrier energy of 59 meV (41). As found for bR, the activation of protein should produce an enhancement of the current response.

5 Conclusions and Perspectives

Overall, our results can be considered promising towards a better understanding of the role of sensing protein topology in the transport mechanisms responsible for charge transfer. The advances in this field are rapid, the results of the application of a combination of different innovative technologies. One example can be cited, where carbon nanotubes have been used to anchor ORs (25). A review of recent such investigations can be found in ref. (60).

The microscopic model described in this chapter is enough flexible to be improved to address questions such as the effect of temperature on transport, the role of dynamics on structure, the parameters responsible for switching the mechanism from tunneling to thermally activated transport, etc.

In conclusion we would like to stress that present results point favorably to the development of a new generation of nanobiosensors within the lab-on-chip strategy.

Acknowledgments

This research is carried out within the bioelectronic olfactory neuron device (BOND) project sponsored by the European Community (EC) within the Seventh Program, grant agreement: 228685-2.

Dr V. Akimov and the INRA-MIG laboratory of Paris are gratefully acknowledged for providing the 3D structures of the OR 17-40.

Drs. E. Pajot, J.-F. Millithaler, N. Jaffrezic-Renault, and C. Pennetta are thanked for useful discussions on the subject.

References

1. Bioelectronic Olfactory Neuron Device (BOND) Collaborative project FP7-NMP-2008-SMALL-2, GA number 228685
2. Pardo M, Sberveglieri G (eds) (2009) Olfaction and electronic nose: proceedings of the 13th international symposium on olfaction and electronic nose. American Institute of Physics, Melville, pp 115–118. ISBN 978-0-7354-0674-2/09/S25.00AIP
3. Quaranta F, Rella R, Siciliano P, Capone S, Epifani M, Vasanelli L et al (1999) A novel gas sensor based on SnO2/Os thin film for the detection of methane at low temperature. Sens Actuators B 58:350–355
4. Gardner JW, Bartlett PN (eds) (1991) Sensor and sensory systems of an electronic nose. Nato ASI series, 212. Kluwer, Dordrecht
5. Göpel W (1998) Chemical imaging: I. Concepts and visions for electronic and bioelectronic noses. Sens Actuators B 52:125–142
6. Göpel W, Ziegler C, Breer H, Schild D, Apfelbach R, Joerges J, Malaka R (1998) Bioelectronic noses: a status report. Part I. Biosens Bioelectron 13:479–493
7. Liu Q, Qu X, Guo H, Chen H, Liu B, Dong S (2006) Olfactory cell-based biosensors: a fist step toward a neurochip of bioelectronic nose. Biosens Bioelectron 22:2195–2201
8. Marrakchi M, Vidic J, Jaffrezic-Renault N, Martelet C, Pajot-Augy E (2007) A new concept of olfactory biosensor based on interdigitated microelectrodes and immobilized yeasts expressing the human receptor OR17-40/*Eur.* Biophys J 36:1015–1018
9. Lefkowitz RJ (2000) The superfamily of heptahelical receptors. Nat Cell Biol 2:E133–E136
10. Palczewski K, Kumasala T, Hori T, Behnke CA, Motoshima H, Fox BA et al (2000) Crystal structure of rhodopsin: a G-coupled protein receptor. Science 289:739–745
11. Buck L, Axel R (1991) A novel multigene family may encode odorant receptors: a molecular basis for odor recognition. Cell 65:175–187
12. Mombaerts P (2004) Odorant receptor gene choice in olfactory sensory neurons: the one receptor-one neuron hypothesis revised. Curr Opin Neurobiol 14:31–36
13. Duchamp-Viret P, Chaput MA, Duchamp A (1999) Odor response properties of rat olfactory receptor neurons. Science 284:2171–2174
14. Buck LB (2004) Olfactory receptors and odor coding in mammals. Nutr Rev 62:S184–S188
15. Crasto C, Singer SM, Shepherd GM (2001) The olfactory receptor family album. Genome Biol 2:reviews 1027-1–reviews 1027-4
16. Jones TD, Reed RR (1989) Golf: an olfactory neuron specific-G protein involved in odorant signal. Science 244:790–795
17. Hou Y, Helali S, Jaffrezic-Renault N, Martelet C, Zhang A, Minic J et al (2006) Immobilization of rhodopsin on a self-assembled monolayer and its specific detection by electrochemical impedance spettroscopy. Biosens Bioelectron 21:1393–1402
18. Hou Y, Jaffrezic-Renault N, Martelet C, Zhang A, Minic J, Gorojankina T et al (2007) A novel detection strategy for odorant molecules based on controlled bioengineering of rat olfactory receptor I7. Biosens Bioelectron 22:1550–1555
19. Levasseur G, Persuy M-A, Grebert D, Remy J-J, Salesse R, Pajot-Augy E (2003) Ligand-specific dose-response in heterologously expressed olfactory receptors. Eur J Biochem 270:2905–2912
20. Pennetta C, Akimov V, Alfinito E, Reggiani L, Gorojankina T, Minic J et al (2006) Towards the realization of nanobiosensors based on G-protein-coupled receptors. In: Kumar CSSR (ed) Nanotechnologies of the life science, vol 4. Wiley, Weinheim, pp 217–240

21. Benilova IV, Minic-Vidic J, Pajot-Augy E, Soldatkin AP, Martelet C, Jaffrezic-Renault N (2008) Electrochemical study of human olfactory receptor OR 17-40 stimulation by odorant in solution. Mater Sci Eng C 28:633–639
22. Vidic J, Grosclaude J, Monnerie K, Persuy M-A, Badonnel K, Baly C et al (2008) On a chip demonstration of a functional role for odorant binding protein in the preservation of olfactory receptor activity at high odorant concentration. Lab Chip 8:678–688
23. Gaillard I, Rouquier S, Giorgi D (2004) Olfactory receptors. Cell Mol Life Sci 61:456–469
24. Vestergaard M, Kerman K, Tamiya E (2007) An overview of label-free electrochemical protein sensors. Sensors 7:3442–3458
25. Yoon H, Lee SH, Kwon OS, Song HS, Oh EH, Park TH et al (2009) Polypyrrole nanotubes conjugated with human olfactory receptors: high-performance transducers for FET-type bioelectronic noses. Angew Chem Int Ed Engl 48:2755–2758
26. Rhodes G (2006) Crystallography made crystal clear: a guide for users of macromolecular models. Academic, New York
27. Berman HM, Westbrook J, Feng Z, Gilliland G, Bhat HWTN, Shindyalov IN, Bourne PE (2000) The protein data bank. Nucleic Acids Res 28:235–242
28. Fiser A, Sali A (2003) ModLoop: automated modeling of loops in protein structures. Bioinformatics 19:2500–2501
29. Sali A, Blundell TL (1994) Comparative protein modelling by satisfaction of spatial restraints. In: Bohr H, Brunak S (eds) Protein structure by distance analysis., pp 64–86
30. Launay G, Téletchéa S, Wade F, Pajot-Augy E, Gibrat J-F, Sanz G (2012) Automatic modeling of mammalian olfactory receptors and docking of odorants. Protein Eng Des Sel 25:377–386
31. Alfinito E, Akimov V, Pennetta C, Reggiani L, Gomila G (2005) Thermal Fluctuations of a GPCR: a two force constant model, AIP Proc. UPON, Conf. Eds. L.Reggiani, C. Pennetta, V. Akimov, E. Alfinito and M. Rosini, 800, pp 381–387
32. Akimov V, Alfinito E, Baussels J, Benilova V, Paramo IC, Errachid A et al (2008) Nanobiosensor based on individual olfactory receptors. Analog Integr Circ Signal Process 57:197–203
33. Alfinito E, Pennetta C, Reggiani L (2008) A network model to correlate conformational change and the impedance spectrum of single proteins. Nanotechnology 19:065202-1–065202-12
34. Alfinito E, Millithaler J-F, Pennetta C, Reggiani L (2009) A nanobiosensor based on olfactory receptor. In: IEEE, 25–28, 10.1109/IWASI.2009.5184762
35. Alfinito E, Reggiani L (2009) Charge transport in bacteriorhodopsin monolayers: the contribution of conformational change to current-voltage characteristics. Europhys Lett 85:68002-1–68002-6
36. Alfinito E, Pennetta C, Reggiani L (2009) Topological change and impedance spectrum of rat olfactory I7: a comparative analysis with bovine rhodopsin and bacteriorhodopsin. J Appl Phys 105:084703-1–084703-6
37. Weiss TF (1997) Cellular biophysics, 2nd edn. MIT, Cambridge
38. Fraufelder H, Wolynes PG, Austin RH (1999) Biological physics. Rev Mod Phys 71:S419–S430
39. Vaidehi N, Floriano WB, Trabanino R, Hall SE, Freddolino P, Choi EJ et al (2002) Prediction of structure and function of G protein-coupled receptors. Proc Natl Acad Sci USA 99:12622–12627
40. Alfinito E, Pennetta C, Reggiani L (2010) Olfactory receptor-based smell nanobiosensors: an overview of theoretical and experimental results. Sens Actuators B 146:554–558
41. Alfinito E, Reggiani L (2010) The role of topology in electrical properties of bR and rat-olfactory receptor. Phys Rev E 81:032902-1–032902-4
42. Alfinito E, Pennetta C, Reggiani L (2010) Smell nanobiosensors: hybrid systems based on the electrical response to odorant capture: theory and experiment. AIP Conf Proc 1137:115–118
43. Alfinito E, Millithaler J-F, Pennetta C, Reggiani L (2010) A single protein based nanobiosensor for odorant recognition. Microelectronics J 41:718–722
44. Alfinito E, Millithaler J-F, Reggiani L, Zine N, Jaffrezic-Renault N (2011) Topological and electrical properties of human 17-40 olfactory receptor for the realization of a nanobiosensor. RCS Adv 1:718–722
45. Alfinito E, Millithaler J-F, Reggiani L (2011) Olfactory receptors for a smell sensor: a comparative study of the electrical responses of rat I7 and human 17-40. Meas Sci Tech 22:124004
46. Alfinito E, Millithaler J-F, Reggiani L (2011) Charge transport in purple membrane monolayers: a sequential tunneling approach. Phys Rev E 83:1–4
47. Atilgan AR, Durell SR, Jernigan RL, Demirel MC, Keskin O, Bahar I (2001) Anisotropy of fluctuation dynamics of proteins with an elastic network model. Biophys J 80:505–515
48. Kim MK, Jernigan RL, Chirikjian GS (2002) Efficient generation of feasible pathways for

protein conformational transitions. Biophys J 83:1620–1630

49. Albert R, Barabasi AL (2002) Statistical mechanics of complex networks. Rev Mod Phys 74:47–97
50. Micheletti C, Carloni P, Maritan A (2004) Accurate and efficient description of protein vibrational dynamics: comparing molecular dynamics and Gaussian models. Proteins 55:635–645
51. Tirion MM (1996) Large amplitude elastic motions in proteins from a single-parameter, atomic analysis. Phys Rev Lett 77:1905–1908
52. Song X (2002) An inhomogeneous model of protein dielectric properties: intrinsic polarizability of amino acids. J Chem Phys 116:9359–9363
53. Churg AK, Weiss RM, Warshel A, Takano T (1983) On the action of cytochrome *c*: correlating geometry changes upon oxidation with activation energy of electron transfer. J Phys Chem 87:1683–1694
54. Warshel A, Parson WW (2001) Dynamics of biochemical and biophysical reactions: insight from computer simulations. Q RevBiophys 34:563–679
55. Hall SE, Floriano WB, Vaidehi N, Goddard WAIII (2004) Predicted 3-D structures for mouse I7 and rat I7 receptors and comparison of predicted odor recognition profiles with experiment. Chem Senses 29:595–616
56. Kolbika BK, Deupi X (2007) Conformational complexity of G-protein-coupled receptors. Trends Pharmacol Sci 28:397–406
57. Miyashita O, Wolynes PG, Onuchic JN (2005) Simple energy landscape model for the kinetics of functional transitions in proteins. J Phys Chem B 109:1959–1969
58. Gardino AK, Villali J, Kivenson A, Lei M, Liu CF, Steindel P et al (2009) Transient non-native hydrogen bonds promote activation of a signaling protein. Cell 139:1109–1118
59. Jin Y, Friedman N, Sheves M, He T, Cahen D (2006) Bacteriorhodopsin (bR) as an electronic conduction medium: current transport through bR-containing monolayers. Proc Natl Acad Sic USA 103:8601–8606
60. Sankaran S, Khot LR, Panigrahi S (2012) Biology and applications of olfactory sensing system: a review. Sens Actuators B 171: 1–17

Chapter 6

A Hit Map-Based Statistical Method to Predict Best Ligands for Orphan Olfactory Receptors: Natural Key Odorants Versus "Lock Picks"

Dietmar Krautwurst and Matthias Kotthoff

Abstract

Smell is a multidimensional chemical sense. It creates a perception of our odorous environment by integrating the information of a plethora of volatile chemicals with other sensory inputs, emotions and memories. We are almost always exposed to odorant mixtures, not just single chemicals. Olfactory processing of complex odorant mixtures, such as coffee or wine, first is decoded at the site of perception by the hundreds of different olfactory receptor types, each residing in the cilia of their olfactory sensory neurons in the nose. Often, only a few odorants from many are essential to determine complex olfactory perception. But merely using the chemical structure of odorants is insufficient to identify and predict characteristic odor qualities and low odor thresholds. An understanding of odorant coding critically depends on knowledge about the interaction of key odorants of biologically relevant odor bouquets with their best cognate receptors. Here, we describe a hit map-based method of correlating the information content of all bioassay-tested odorants with their cognate odorant-receptor frequency in four phylogenetic subsets of human olfactory/chemosensory receptors.

Key words Olfactory receptors, Chemosensory receptors, Key food odorants, Body odorants, Phylogenetic dendrogram, Hit map, Cognate odorant-receptor frequency

1 Introduction

Mammals have an exquisite sense of smell. The high discriminatory power associated with this sense arises from up to three unrelated gene families in humans, which code for rhodopsin-like G protein-coupled seven transmembrane helix receptors (GPCR): odorant receptors (OR) of the main olfactory epithelium (MOE) (1, 2), vomeronasal type-1 receptors (VN1R) (3–5), and trace amine-associate receptors (TAAR) (6–8). The detection of a multitude of volatile compounds is thought to be achieved predominantly by the large family of about 390 putative, functional, human ORs (9–15). These ORs fall into two major phylogenetic clades, the class I and II ORs (12, 16). OR repertoires appear to

Chiquito J. Crasto (ed.), *Olfactory Receptors: Methods and Protocols*, Methods in Molecular Biology, vol. 1003,
DOI 10.1007/978-1-62703-377-0_6, © Springer Science+Business Media, LLC 2013

be shaped mainly by numerous gene duplication events and a lack of negative selection. This has resulted in a large number of olfactory receptor pseudogenes (>50%) (13, 17–19). Moreover, human OR genes carry more than 1,000 single nucleotide polymorphisms (SNPs) within their coding regions (~2.5–3.0 SNPs per OR) (9, 19, 20) (D.K., unpublished observations, but see Note 1), which is about twofold compared with non-olfactory GPCRs. This suggests that the human olfactory receptor space is designed to detect a vast diversity of volatile stimuli. Many of today's synthetic odorants found in, for instance, perfumes or aromas may potently activate OR ("lock picks"). But these lack any evolutionary and biological relevance.

OR repertoires of any species may reflect an evolutionary adaptation to its specific environmental (ecological and social) requirements (21). Testing volatile molecules versus ORs in bioassays that are important carriers of information for intra-species communication by pheromones (e.g., body odors, BO) might provide interesting insights into OR-odor interactions, interspecies signaling by allomones or kairomones, or the identification and evaluation of food sources via key food odorants (KFO). The latter are defined as odorants that occur naturally in foods at concentrations above their odor thresholds. More strictly defined—by establishing minimal aroma recombinates and performing omission experiments of single aroma components—KFOs have been found to be necessary and sufficient to determine the odor of the foods in which they are present (22). About 8,000 volatiles have been identified just in food, however, the total number of KFO has been estimated at a number below 400 (23). This appears to be close to the number of functional human OR genes.

In the past, screening experiments of hundreds of OR versus hundreds of mostly non-key food or synthetic odorants resulted in few best matches: only as few as 23 (~6%) human ORs could be assigned to specific odorants. Additionally, volatile agonists have been described for one (TAAR1) (see Note 2), and all VN1Rs. Yet, information on the full odorant spectra of the majority of human OR is lacking.

Here we report a hit map-based method to correlate the information content of bioassay-tested odorants with the frequency of cognate odorant-receptor pairs. For all the ORs, this correlation shows that cognate key odorant-receptor pairs were up to six times more frequent than non-key odorant-receptor pairs, despite the fact that key odorants (KFO, BO) were underrepresented among all the odorants tested in bioassay-based screening experiments published. This correlation led us to hypothesize that an odorant space defined by the most abundant KFO and BO may largely overlap as best cognate matches with the reminiscent functional human OR space.

We propose, as a first step, to systematically screen at least a significant portion of an entire olfactory receptor repertoire against selected, biologically relevant odorants, e.g., putative human pheromones (24–30), or key food odorants (30–32), in order to identify the best cognate odorant-receptor pairs. With this study we aim to provide a rough toolbox for considering parameters for future screening approaches of odorants versus receptors.

2 Materials

2.1 Databases and Search Engines

1. NCBI and HORDE (see Chapter 2) were used as databases for the retrieval of genetic information on human and mouse OR, TAAR, and human VN1R (9, 33).
2. The phylogenetic reconstruction of OR was performed using the software *MEGA* version 5 (34).
3. Amino acid identity of mouse and human receptors was calculated with CLCbio software (35).
4. To retrieve publications on odorants used in bioassay-based screenings versus OR, we used the search engines PubMed and SciFinder (36, 37).

2.2 Odorants, Nomenclature, Exclusion Criteria, and Categories

1. In several different publications, the authors have used diverse nomenclature and trivia to refer to the odorants used in experiments. We adopt a normalized system of identifying odorants that have been subject to OR research, by using a homogenous and systematic nomenclature, where applicable (e.g., terpenes may have very complex systematic name, but are clearly assignable by their trivial names). This was also necessary in order to avoid multiple names for the same substance.
2. Complex mixtures of odorants that have been used in screening approaches, such as oil of turpentine and sweet orange oil as components of the frequently used "Henkel 100" odorant-mix, have been excluded from our analysis.
3. The odorants, used in our study, are sorted into the three categories are (in alphabetical order):

   KFO: (−)-Carveol, (−)-Carvone, (−)-Citronellal, (−)-Citronellol, (−)-Fenchone, (+)-Camphor, (+)-Carvone, (+)-Fenchone, (*E*)-2-nonenal, (*R*)-(−)-Myrtenal, (*R*)-(+)-Limonene, (*S*)-(−)-Limonene, 1,4-Cineole, 1,8-Cineole, 1-octen-3-ol, 1-Octen-3-one, 1-Penten-3-one, 2,3-Butandione, 2, 3-Diethyl-5-methylpyrazine, 2,3-Dimethylpyrazine, 2-Butanone, 2-Decenal, 2-Ethyl pyrazine, 2-Furfurylthiol, 2-Heptanone, 2-Hexanone, 2-Isobutyl-3-methoxypyrazine, 2-Isopropyl-3-methoxypyrazine, 2-Methoxy-4-methyl phenol, 2-Methoxyphenol, 2-Methyl propanal, 2-Methyl pyrazine, 2-Methylbutanoic acid, 2-Methylpropanoic acid, 2-Methylpyrazine, 2-Octanone, 3-Hydroxy-2-butanon, 3-Mercapto-1-hexanol, 3-Methyl butanoic acid, 3-*sec*-Butyl-2-methoxypyrazine, a-Citral, Androstadienone, Androstenol,

Androstenone, Anethole, Anisaldehyde, Butanal, Butanoic acid, 1-Butanol, Butyl acetate, Camphor, Cinnamaldehyde, Cinnamyl alcohol, *cis*-3-Hexen-1-ol, Citral, Coumarin, Damascone, δ-Decalactone, Decanal, Decanoic acid, Diallyl sulfide, Dimethyl sulfide, Dodecanal, *E,E*-2,4-Decadienal, *E,E*-2,4-Nonadienal, *E,Z*-2,6-Nonadienal, *E*-4-Decenal, Ethyl hexanoate, Ethyl octanoate, Eucalyptol, Eugenol, Farnesene, Farnesol, Fenchone, Furaneol, γ-Decalacton, Geraniol, γ-Octalactone, Heptanal, Heptanoic acid, Hexanal, Hexanoic acid, 1-Hexanol, Hexyl acetate, Hexyl butyrate, Indole, Isoeugenol, Linalool, Maltol, Menthol, Menthone, Methanethiol, Methional, Methionol, Nerol, Nerolidol, Nonanal, Octanal, Octanoic acid, *p*-Cymene, Pentanoic acid, Phenyl acetaldehyde, Phenylethyl acetate, Piperonal, Propanoic acid, Pyrrolidine, 4-(4-Hydroxyphenyl) butan-2-one, β-Damascenone, β-Ionone, β-Phenylethyl amine, Terpineol, Terpinyl acetate, Thymol, Trimethylamine, Trimethylpyrazine, Undecanal, Vanillic acid, Vanillin, *Z*-4-Decenal, *Z*-7-Decanal, and α-Ionone.

BO: (−)-Citronellol, (+)-Citronellol, (*E*)-2-nonenal, 1-octen-3-ol, 3-Mercapto-1-hexanol, 3-Methyl butanoic acid, Acetophenone, Androstadienone, Androstenone, Butanal, Butanoic acid, Camphor, Decanal, Decanoic acid, Dodecanal, 1-Dodecanol, Geraniol, Hexanoic acid, Linalool, Menthol, Nonanoic acid, Octanal, Octanoic acid, *p*-Cymene, Phenylethyl alcohol, Propanoic acid, Terpineol, and Undecanol.

Others: (−)-2-Phenylbutanoic acid, (−)-Carvyl acetate, (+)-2-Phenylbutanoic acid, (+)-Citronellal, (+)-Dihydrocarvone, 1,2-Dichlorethylen, 1-Pyrrolidine, 2,3-Diethylpyrazine, 2, 3-Hexanedione, 2,5-Dimethylpyrazine, 2-Bromo-Octanoic acid, 2-Bromo-pentanoic acid, 2-Coumaranone, 2-Ethyl-1-hexanol, 2-Ethylfenchol, 2-Methoxy-3-methyl pyrazine, 2-Methyl-4-propyl-1,3-oxathiane, 2-Methylbutyl amine, 2-Nonanol, 2-Nonanone, 3,4-Hexanedione, 3-Heptanone, 3-Methyl pentanoic acid, 3-Methylbutyl acetate, 3-Octanone, 4-Chromanone, 4-Heptanone, 4-Hydroxycoumarin, Acetyl muscone, Allyl benzene, Allyl heptanoate, Allyl phenylacetate, Ambroxan, Amyl butyrate, Amyl cinnamaldehyde, Amyl hexanoate, Anthoxan, Belambra, Benzene, Benzophenone, Benzothiazol, Benzyl acetate, Benzyl alcohol, Benzyl benzoate, Benzyl salicylate, Boisambrene forte, Bourgeonal, Brahmanol, Bromobutanoic acid, Butanethiol, Butanoic diacid, Butyl butyrylactate, Butyl formate, Cedryl acetate, Citral dimethyl acetate, Citralva, Citrathal, Citronellic acid, Citronellyl nitril, Citrusal, Cyclamen aldeyde, Cyclohexanone, Cyclohexyl salicylate, Cyclohexylamine, Cyclosal,

Cyclovertal, Decanol, Dihydro myrcene, Dihydrojasmone, Dimedone, Dimethylacetal, Dimetol, Diphenyl ether, DMBCA ((2-methyl-1-phenylpropan-2-yl)-acetate), *E,E*-2,4-Dodecadienal, Epiandrostenone, Ethyl decanoate, Ethyl heptanoate, Ethyl nonanoate, Ethyl vanillin, Ethyl-2-methyl propanoate, Ethylene brassylate, Ethylfruitat (Fructone), Eugenol acetate, Eugenol methyl ether, Evernyl, Fixolide np, Floramat, Foliaver, Galaxolide, Geranonitrile, Geranyl acetate, Hedione, Helional, Heliotropyl acetone, Heptanoic diacid, Heptanol, Heptyl acetate, Herbavert, Hexanoic diacid, Hexyl cinnamaldehyde, Irotyl, Isoamylamine, Isobornyl acetate, Isononyl acetate, Jasmacyclat, Ligustral, Lilial, Lyral, Mathyl hexanoate, Melusat, Methyl decanoate, Methyl heptanoate, Methyl nonanoate, Methyl octanoate, Methyl salicylate, Methylpentanal, Methyl-phenyl pentanal, Muscone, Neobergamate, Nerylacetate, *N*-Methylpiperidine, Nonanoic diacid, Nonanethiol, Nonanol, Octanethiol, Octanoic diacid, Octanol, Octyl acetate, OTBCA (2-*tert*-Butylcyclohexylacetat), Pentadecalactone, Pentanal, Pentanol, Pentanthiol, Peranat, Phenirat, Phenyl acetate, Prenyl acetate, Propidyl acetate, Propylene glycol, PTBCA (4-*tert*-Butylcyclohexylacetat), Pyrazine, Pyridine, Quinoline, Romilat, Sandalore, Sandelice, S-methyl thiobutanoate, Styrlyl acetate, Testosterone, Thiazol, Traseolide, Triethylamine, Tridecanal, Trifemal, Troenan, Tyramine, and Verdoxan.

3 Methods

3.1 Evolutionary Relationships of Human Chemosensory Receptors

1. Phylogenetic and molecular evolutionary analyses were conducted using the software *MEGA* version 5.
2. The evolutionary history of 391 human OR, 6 TAAR and 5 VN1R was inferred using the neighbor-joining method (38), followed by 500 bootstrap replicates (39).
3. There were a total of 283 positions in the final dataset.
4. The result is displayed as a linear phylogenetic dendrogram on the left hand side of Fig. 1.

3.2 Sorting by Information Content (Environmental Appearance) of All Odorants Used in Bioassay-Based Screening Experiments Versus OR

1. To organize all 285 published odorants according to their environmental appearance, we sorted them into three major groups: a total of 121 key food odorants (KFO), 28 body odorants (BO), and 158 others without environmentally relevant group assignment. 22 out of 28 odorants in the BO category overlap with odorants from the KFO list.
2. As KFO, all odorants were classified that have been

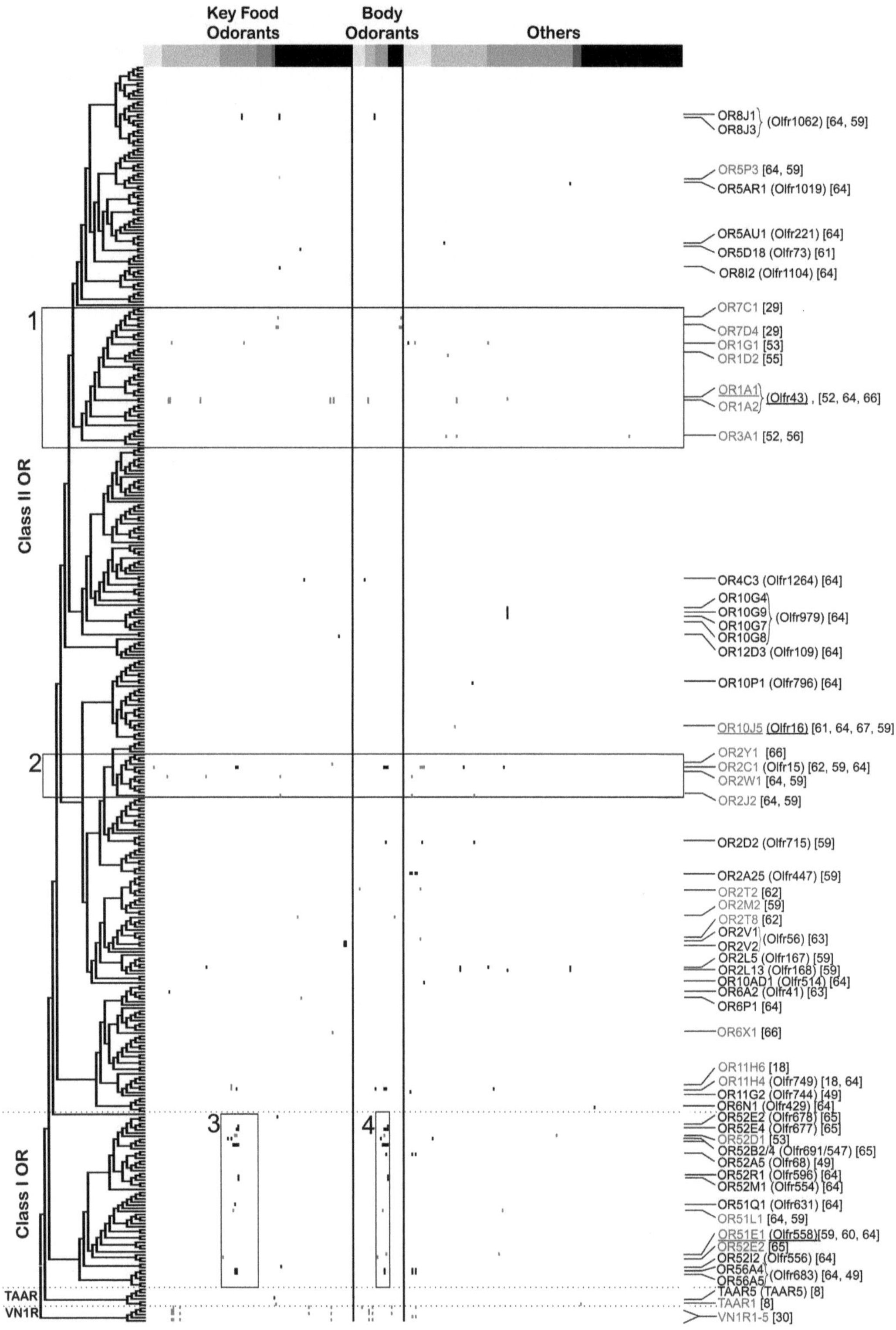

Fig. 1 Hit map of cognate odorant-receptor pairs. On the *y*-axis, the phylogenetic relationships of all functional human olfactory receptors (Class I- and Class II-OR, TAAR, and VN1R, separated by *dashed lines*) are displayed. The *x*-axis gives all odorants that have been used in screening experiments, so far. The odorants are sorted by

(a) Described to have an Odorant Activity Value (OAV) ≥1 (40, 41), which is the ratio of the odorant-concentration in a certain food and the detection threshold of this compound

$$\left[\mathrm{OAV} = \left(\frac{c(\mathrm{odorant})}{c(\mathrm{detection\ threshold}}\right)\right], \text{ or}$$

(b) Identified through Gas Chromatography-Olfactometry (GC-O) and Aroma Extract Dilution Analyses (AEDA) to contribute to the aroma of a certain food (22, 42).

3. All chemicals from the BO list have been described as components of human body odor and sweat, independent from odor threshold and volatility (43–47). Some of the described body odorants, however, are of synthetic origin only. They are likely to be remnants from cosmetics used by the tested individuals. These odorants (e.g. Galaxolide, Lilial, and Hedione) were not considered body odorants.
4. All remaining volatiles not in the BO or KFO list are classified as "Others".

3.3 Sorting of Screening Odorants by Functional Groups or Physico-Chemical Characteristics

1. We further sub-grouped the odorants by their functional groups. For chemicals with more than one functional group were identified by the functional group which provided a dominant odor-signature. Examples include thiols, which are responsible for potent odors, amines which presumably target specific receptors (TAAR), or pyrazines, which are products of the Maillard reaction, and are commonly found in heated food.
2. Within each function group sub-class, the chemicals were arranged by carbon chain length (homologous series) or molecular size, whenever applicable.

3.4 Hit-Mapping Functional Information from De-orphaned Mouse OR to Putative Isofunctional Human Orthologs

1. More than half of all human ORs have closest relatives in the mouse OR repertoires with >80% sequence similarity (48), and are likely to be orthologs with an evolutionarily conserved function (49). We included in our analysis mouse orthologs that had responded previously to odorants in any bioassay-based screening experiment (see Note 2).
2. In order to identify such orthologs, we performed a ClustalW alignment with 1,119 mouse and 391 human ORs, as well as 16 mouse and 6 human TAARs using the CLCbio Software.

Fig. 1 (continued) means of chemical characteristics, such as functional groups, indicated by *greyscale bars* as follows: □ alcohols/thiols; □ aldehydes/ketones; □ acids/esters/lactones; □ pyrazines; ■ amines; ■ terpenes/others. Within these groups they are sorted by other molecular descriptors, such as carbon chain length. Published cognate odorant-receptor pairs are shown as *gray* and *black spots* (for tested human receptors and inferred human orthologs to tested mouse receptors, respectively). On the *right hand side*, receptor names (mouse OR in parentheses) are indicated with the corresponding literature given in *brackets* (8, 18, 29, 30, 50, 51, 54–57, 60, 62–69). Human and mouse orthologs that have been described with the same odorant ligand are *underlined*

3. Mouse ORs for which at least one odorant had been identified and which have human orthologs, and accounted the odorant ligand information for the human OR in the hit map were identified.
4. This analysis of orthologous receptors revealed information on putative ligands for an additional 40 human ORs and 1 TAAR (see Notes 2, 3, and 7).
5. Isofunctional orthologs have been shown previously for some OR, e.g., human OR1A1 and mouse Olfr43, which both responded to (−)-Citronellal and share 82.75% sequence identity on the amino-acid level (50) (see Notes 2 and 7).

3.5 Odorant-Receptor Hit-Mapping Reveals a Clustered Pattern of Receptor Activating Odorants to Phylogenetic Receptor Subsets

1. An *in silico* analysis of the phylogenetic relationships of human OR, together with a mapping of all published activating odorants (see Notes 3–5) suggests, at first glance, that ORs from distinct phylogenetic clades may be activated by odorants with certain physico-chemical characteristics, such as functional groups (Fig. 1).
2. Class-I OR is the oldest evolutionary OR class, which is also found in fish, suggesting rather soluble odorant ligands with low cLogP values. Several groups independently found carboxylic acids and esters as agonists for human class-I ORs (Fig. 1, Boxes 3 and 4). This corroborates the results of an earlier report describing exclusively mono- and dicarboxylic acids as agonists for mouse class-I ORs (51).
3. Five published de-orphaned human OR (OR1A1, OR1A2, OR3A1, OR1D2, OR1G1) are encoded by genes from one of the best characterized human OR gene clusters on chromosome 17@3.1 (52, 53). Interestingly, these OR share their best agonists, aliphatic and aromatic, fruity, and floral aldehydes (50, 54–57). Other than that, some subfamilies of Class-II ORs appear to show a rather broad spectrum of chemically versatile odorants (Fig. 1 Boxes 1 and 2), whereas VN1R seem to have a more narrowly tuned ligand profile. Several groups have discussed the existence of receptor specialists, having a narrowly tuned ligand profile versus those with a rather broadly tuned odorant ligand spectrum (52, 54, 58, 59) (but see Note 6).

3.6 Parameters and Statistical Evaluation of the Hit Map Data

1. To allow for a statistical estimation of the relation between phylogenetically related ORs and their cognate odorants from the three categories, we have defined a set of descriptive parameters.
2. *Total odorants*: KFO, BO and Others, are displayed on the *x*-axis of the hit map. All these odorants have been previously used in bioassay-based screening experiments on chemoreceptors from the three categories (OR, TAAR, VN1R), albeit not the entire odorant collection was tested on each receptor, or each receptor subset.

3. *Hits*: are all functional odorant/OR combinations.
4. *Bioassay active odorants*: is the number of odorants that could activate at least one receptor. In our study, this number is smaller than the number of "Total odorants," since not all odorants were tested versus all receptors.
5. *Cognate Olfactory Receptor Index* (*CORI*), is "Hits" over "bioassay-active odorants", and is a relative measure for the number of olfactory receptors, that could be functionally identified per odorant:

$$CORI = \frac{n(\text{Hits})}{n(\text{Bioassay} - \text{active odorants})}$$

6. *Hit Rate*, is "Hits" per "Total odorants," and is the probability for cognate odorant-receptor pairs in bioassay-based screening approaches. This number may become larger than 1 because of the combinatorial odorant coding of OR:

$$Hite\ rate = \frac{n(\text{Hits})}{n(\text{Total odorants})}$$

7. *Cognate Olfactory Receptor Frequency*, *CORF*, which is the product of "CORI" and "Hit rate":

$$CORF = \text{CORI} \times \text{Hit rate} = \frac{n(\text{Hits})^2}{n(\text{Bioassay} - \text{active odorants}) \times n(\text{Total odorants})}$$

8. We extracted the numbers for these parameters from the hit map in Fig. 1, and summarized them for the three odorant categories and all receptor subsets in Table 1.
9. The CORF suggests an enhanced chance of identifying cognate odorant-receptor pairs by testing an entire receptor repertoire with odorants from the KFO (CORF = 1.78) and BO (CORF = 4.32) groups, compared to other odorants (CORF = 0.71). This becomes even more evident when testing biologically relevant odorants versus specific receptor subsets, such as class-I OR, with CORF values of 0.50, 1.93, and 0.11 for KFO, BO and Others, respectively, or VN1R (see Table 1).

4 Notes

1. According to the most recent literature, 413 is the actual number of intact OR loci, with on average about 10 SNPs per locus (70).

Table 1
Parameters and statistical evaluation of the hit map data

	KFO	BO	Others
Total odorants	121	28	158
All OR (*n* = 402)			
Bioassay-active odorants	32	16	30
Hits	83	44	58
CORF	1.78	4.32	0.71
Class-I OR (*n* = 56)			
Bioassay-active odorants	8	6	6
Hits	22	18	10
CORF	0.50	1.93	0.11
Class-II OR (*n* = 335)			
Bioassay-active odorants	24	10	24
Hits	44	17	43
CORF	0.67	1.03	0.49
TAAR (*n* = 6)			
Bioassay-active odorants	2	2	1
Hits	2	2	1
CORF	0.02	0.07	0.01
VN1R (*n* = 5)			
Bioassay-active odorants	5	4	2
Hits	15	9	3
CORF	0.37	0.72	0.03

KFO key food odorants, *BO* body odorants, *OR* odorant receptors, *TAAR* trace amine-associated receptors, *VN1R* vomeronasal type-I receptors. Numbers for these parameters are from the hit map in Fig. 1. *CORF* cognate odorant-receptor frequency

2. In the present study, we included ligand information from de-orphaned mouse orthologs of human ORs and TAARs, assuming that the human receptors would respond to the same odorants (e.g. trimethylamine and mouse Taar5). A most recent study has now proven our approach to be valid for TAAR5, by demonstrating that TAAR5 indeed responded to trimethylamine (71).
3. Taken together, the sparse information on cognate human OR/odorant pairs that has been published so far is utterly heterogenous.

This is with respect to the different cell-based bioassays employed, the different signal transduction G proteins involved, and the number of tested ORs. The maximum number of human OR in a screen versus selected odorants amounted to 245 (60).

4. Also, the number of tested odorants, and the different odorant concentrations tested varies significantly. This ranges from single odorant concentration amplitudes obtained from single OR type-expressing cells to EC_{50} values for homologous series of odorants on OR families or partial OR libraries.
5. Moreover, it has not been demonstrated, so far, whether all recombinant ORs express equally well at the plasma membrane of test cell systems. Thus, the quality of the data sets used for this study had to be quite diverse.
6. For the above mentioned reasons, care has to be taken when interpreting odorants as the best cognate ligands defining the odorant specificity of single OR types. Additionally, addressing OR as specialists or generalist may not be appropriate, yet.
7. During revision of this manuscript, Duan et al. published another murine olfactory receptor, Olfr1509, to selectively respond to a copper-(methylthio)methanethiol complex, the latter substance being a component of male mouse urine (61). However, this compound and other tested homologs are not known to play a role as KFO, and are not described to be a component of human BO. So far, there are no odorant data for the closest human relative, OR4E2 (86% amino acid identity). Recently, activating odorants have been reported for 18 human and mouse OR orthologs (72).

References

1. Buck L, Axel R (1991) A novel multigene family may encode odorant receptors: a molecular basis for odor recognition. Cell 65(1): 175–187
2. Levy NS, Bakalyar HA, Reed RR (1991) Signal transduction in olfactory neurons. J Steroid Biochem Mol Biol 39(4B):633–637
3. Dulac C, Axel R (1995) A novel family of genes encoding putative pheromone receptors in mammals. Cell 83(2):195–206
4. Rodriguez I et al (2000) A putative pheromone receptor gene expressed in human olfactory mucosa. Nat Genet 26(1):18–19
5. Rodriguez I, Mombaerts P (2002) Novel human vomeronasal receptor-like genes reveal species-specific families. Curr Biol 12(12):R409–R411
6. Lewin AH (2006) Receptors of mammalian trace amines. AAPS J 8(1):E138–E145
7. Liberles SD (2009) Trace amine-associated receptors are olfactory receptors in vertebrates. Ann N Y Acad Sci 1170:168–172
8. Liberles SD, Buck LB (2006) A second class of chemosensory receptors in the olfactory epithelium. Nature 442(7103):645–650
9. HORDE (2011) The Human Olfactory Receptor Data Exploratorium (HORDE), The Weizmann Institute, http://bioportal.weizmann.ac.il/HORDE/
10. Malnic B, Godfrey PA, Buck LB (2004) The human olfactory receptor gene family. Proc Natl Acad Sci USA 101(8):2584–2589
11. Menashe I, Aloni R, Lancet D (2006) A probabilistic classifier for olfactory receptor pseudogenes. BMC Bioinformatics 7:393
12. Niimura Y, Nei M (2003) Evolution of olfactory receptor genes in the human genome. Proc Natl Acad Sci USA 100(21):12235–12240

13. Olender T et al (2004) The olfactory receptor universe–from whole genome analysis to structure and evolution. Genet Mol Res 3(4):545–553
14. Olender T, Lancet D, Nebert DW (2008) Update on the olfactory receptor (OR) gene superfamily. Hum Genomics 3(1):87–97
15. Zozulya S, Echeverri F, Nguyen T (2001) The human olfactory receptor repertoire. Genome Biol 2(6):RESEARCH0018
16. Glusman G et al (2000) The olfactory receptor gene superfamily: data mining, classification, and nomenclature. Mamm Genome 11(11): 1016–1023
17. Gilad Y, Lancet D (2003) Population differences in the human functional olfactory repertoire. Mol Biol Evol 20(3):307–314
18. Menashe I et al (2007) Genetic elucidation of human hyperosmia to isovaleric acid. PLoS Biol 5(11):2462–2468
19. Menashe I et al (2003) Different noses for different people. Nat Genet 34(2):143–144
20. Sharon D et al (2000) Identification and characterization of coding single-nucleotide polymorphisms within a human olfactory receptor gene cluster. Gene 260(1–2):87–94
21. Hayden S et al (2010) Ecological adaptation determines functional mammalian olfactory subgenomes. Genome Res 20(1):1–9
22. Grosch W (2001) Evaluation of the key odorants of foods by dilution experiments, aroma models and omission. Chem Senses 26(5): 533–545
23. Grosch W (2000) Specificity of the human nose in perceiving food odorants. In: Frontiers of flavour science, Ninth Weurman flavour research symposium. Deutsche Forschungsanstalt fuer Lebensmittelchemie, Garching
24. Wyart C et al (2007) Smelling a single component of male sweat alters levels of cortisol in women. J Neurosci 27(6):1261–1265
25. Doucet S et al (2007) The "smellscape" of mother's breast: effects of odor masking and selective unmasking on neonatal arousal, oral, and visual responses. Dev Psychobiol 49(2):129–138
26. Boulkroune N et al (2007) Repetitive olfactory exposure to the biologically significant steroid androstadienone causes a hedonic shift and gender dimorphic changes in olfactory-evoked potentials. Neuropsychopharmacology 24:24
27. Miller MC (2006) Human pheromones. Harv Ment Health Lett 23(5):8
28. Laska M, Wieser A, Salazar LT (2006) Sex-specific differences in olfactory sensitivity for putative human pheromones in nonhuman primates. J Comp Psychol 120(2):106–112
29. Keller A et al (2007) Genetic variation in a human odorant receptor alters odour perception. Nature 449(7161):468–472
30. Shirokova E et al (2008) The human vomeronasal type-1 receptor family–detection of volatiles and cAMP signaling in HeLa/Olf cells. FASEB J 22(5):1416–1425
31. Krautwurst D (2008) Human olfactory receptor families and their odorants. Chem Biodivers 5(6):842–852
32. Shirokova E et al (2005) Identification of specific ligands for orphan olfactory receptors. G protein-dependent agonism and antagonism of odorants. J Biol Chem 280(12): 11807–11815
33. NCBI (2011) National Center for Biotechnology Information, Gene search tool, Available from: http://www.ncbi.nlm.nih.gov/gene/
34. Tamura K et al (2011) MEGA5: molecular evolutionary genetics analysis using maximum likelihood, evolutionary distance, and maximum parsimony methods. Mol Biol Evol 28(10):2731–2739
35. Knudsen B et al. (2011) CLC Main Workbench 6.5, CLCbio
36. PubMed (2011) PubMed health, Available from: http://www.ncbi.nlm.nih.gov/pubmedhealth/
37. SciFinder (2009) Chemistry research tool, Available from: https://scifinder.cas.org
38. Saitou N, Nei M (1987) The neighbor-joining method: a new method for reconstructing phylogenetic trees. Mol Biol Evol 4(4): 406–425
39. Felsenstein J (1985) Confidence limits on phylogenies: an approach using the bootstrap. Evolution 39:783–791
40. Acree TE et al (1984) Food Chem. 14(4): 273–286
41. Rothe M, Thomas B (1963) Z. Lebensm.-Unters. Forsch. 119:302–310
42. Schieberle P (1995) New developments on methods for analysis of volatile flavor compounds and their precursors. In: Gaonkar AG (ed) Characterization of food-emerging methods. Elsevier, Amsterdam, pp 403–433
43. Gallagher M et al (2008) Analyses of volatile organic compounds from human skin. Br J Dermatol 159(4):780–791
44. Hasegawa Y, Yabuki M, Matsukane M (2004) Identification of new odoriferous compounds in human axillary sweat. Chem Biodivers 1(12):2042–2050
45. Natsch A et al (2006) A broad diversity of volatile carboxylic acids, released by a bacterial

aminoacylase from axilla secretions, as candidate molecules for the determination of human-body odor type. Chem Biodivers 3(1):1–20

46. Natsch A et al (2003) A specific bacterial aminoacylase cleaves odorant precursors secreted in the human axilla. J Biol Chem 278(8): 5718–5727
47. Natsch A, Schmid J, Flachsmann F (2004) Identification of odoriferous sulfanylalkanols in human axilla secretions and their formation through cleavage of cysteine precursors by a C-S lyase isolated from axilla bacteria. Chem Biodivers 1(7):1058–1072
48. Zhang X, Firestein S (2002) The olfactory receptor gene superfamily of the mouse. Nat Neurosci 5(2):124–133
49. Tatusov RL, Koonin EV, Lipman DJ (1997) A genomic perspective on protein families. Science 278(5338):631–637
50. Schmiedeberg K et al (2007) Structural determinants of odorant recognition by the human olfactory receptors OR1A1 and OR1A2. J Struct Biol 159(3):400–412
51. Malnic B et al (1999) Combinatorial receptor codes for odors. Cell 96(5):713–723
52. Lapidot M et al (2001) Mouse-human orthology relationships in an olfactory receptor gene cluster. Genomics 71(3):296–306
53. Glusman G et al (2001) The complete human olfactory subgenome. Genome Res 11(5): 685–702
54. Jacquier V, Pick H, Vogel H (2006) Characterization of an extended receptive ligand repertoire of the human olfactory receptor OR17-40 comprising structurally related compounds. J Neurochem 97(2):537–544
55. Sanz G et al (2005) Comparison of odorant specificity of two human olfactory receptors from different phylogenetic classes and evidence for antagonism. Chem Senses 30(1):69–80
56. Spehr M et al (2003) Identification of a testicular odorant receptor mediating human sperm chemotaxis. Science 299(5615): 2054–2058
57. Wetzel CH et al (1999) Specificity and sensitivity of a human olfactory receptor functionally expressed in human embryonic kidney 293 cells and Xenopus Laevis oocytes. J Neurosci 19(17):7426–7433
58. Grosmaitre X et al (2009) SR1, a mouse odorant receptor with an unusually broad response profile. J Neurosci 29(46):14545–14552
59. Grus WE, Zhang J (2008) Distinct evolutionary patterns between chemoreceptors of 2 vertebrate olfactory systems and the differential tuning hypothesis. Mol Biol Evol 25(8):1593–1601
60. Saito H et al (2009) Odor coding by a mammalian receptor repertoire. Sci Signal 2(60):ra9
61. Duan X et al (2012) Crucial role of copper in detection of metal-coordinating odorants. Proc Natl Acad Sci USA 109(9):3492–3497
62. Fujita Y et al (2007) Deorphanization of Dresden G protein-coupled receptor for an odorant receptor. J Recept Signal Transduct Res 27(4):323–334
63. Kajiya K et al (2001) Molecular bases of odor discrimination: reconstitution of olfactory receptors that recognize overlapping sets of odorants. J Neurosci 21(16):6018–6025
64. Kotthoff M, Schieberle P, Krautwurst D(2010) Human olfactory receptors for thiol key food odorants. In: 9th Wartburg symposium on flavor chemistry & biology. Eisenach, Chair for Food Chemistry, Technische Universitaet Muenchen
65. Krautwurst D, Yau KW, Reed RR (1998) Identification of ligands for olfactory receptors by functional expression of a receptor library. Cell 95(7):917–926
66. Li YR, Matsunami H (2011) Activation state of the m3 muscarinic acetylcholine receptor modulates Mammalian odorant receptor signaling. Sci Signal 4(155):ra1
67. Saito H et al (2004) RTP family members induce functional expression of mammalian odorant receptors. Cell 119(5):679–691
68. Schmiedeberg K (2004) Molekulare klonierung, charakterisierung und struktur-funktions-beziehungen von olfaktorischen rezeptoren für schlüsselaromastoffe, In: Mathematisch-naturwissenschaftliche fakultät 2004, Universität Potsdam, Potsdam
69. Touhara K et al (1999) Functional identification and reconstitution of an odorant receptor in single olfactory neurons. Proc Natl Acad Sci USA 96(7):4040–4045
70. Olender S et al (2012) BMC Genomics 13:414
71. Wallrabenstein I (2013) PLoS One 8(2):e54950
72. Adipietro KA et al (2012) PLoS 8(7):p. e1002821

Part III

Expression Systems for Olfactory Receptors

Chapter 7

A Robust, Rapid, and Simple Method of Producing Olfactory Receptors Using Commercial *E. coli* Cell-Free Systems

Karolina Corin, Brian Cook, and Shuguang Zhang

Abstract

The first bottleneck in olfactory receptor (OR) studies is producing sufficient quantities of soluble, functional, and stable receptors. Commercial cell-free in vitro translation systems can be used to produce milligrams of soluble and functional receptors within several hours directly from plasmid DNA. The receptors can be purified using immunoaffinity chromatography and gel filtration, and can be analyzed using gel electrophoresis and with other standard techniques.

Key words In vitro translation, Olfactory receptors, Odorant ligand-binding, Microscale thermophoresis

1 Introduction

The molecular basis of olfaction is poorly understood, primarily due to the difficulty of expressing sufficient quantities of soluble and functional olfactory receptors (ORs). ORs belong to the G protein-coupled receptor family, which is characterized by seven transmembrane helical segments arranged in a barrel-like conformation. These transmembrane regions, which include the receptor binding pocket, can cause problems with protein expression, and make it difficult to functionally stabilize the receptors outside of their native membrane environment. Receptor production in eukaryotic or bacterial cells frequently encounters problems such as low yields, cell toxicity, protein degradation, protein inhomogeneity, and aggregation in internal compartments or inclusion bodies (1–5). Cell-free in vitro translation is an alternative method allowing for rapid, cost-effective, high-yield protein expression (6–11).

Cell-free expression is an established technology for producing soluble proteins. This can be adapted for membrane proteins by including an appropriate detergent in the reaction mixture (7–11) (Fig. 1). Indeed, by using the optimal detergent, it is possible to use cell-free systems to rapidly produce milligram quantities of receptors within several hours directly from plasmid DNA.

Chiquito J. Crasto (ed.), *Olfactory Receptors: Methods and Protocols*, Methods in Molecular Biology, vol. 1003,
DOI 10.1007/978-1-62703-377-0_7, © Springer Science+Business Media, LLC 2013

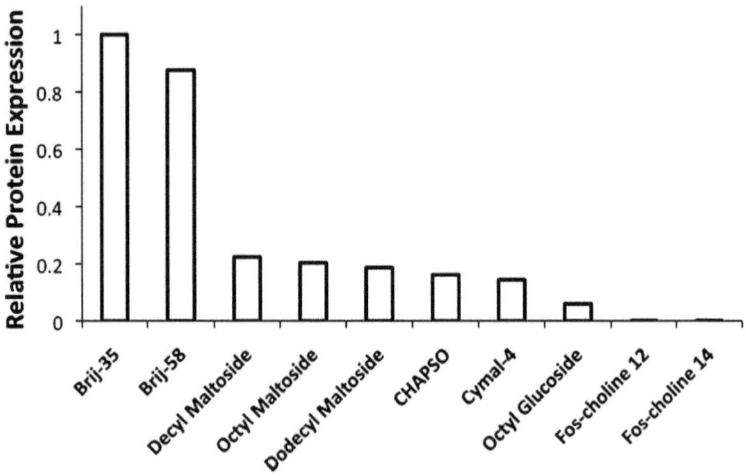

Fig. 1 Detergent screen showing the relative expression of hOR17-210 in ten different detergents. The Brij-35 and Brij-58 detergents yielded four to five times more protein than the next best detergent that was tested (8)

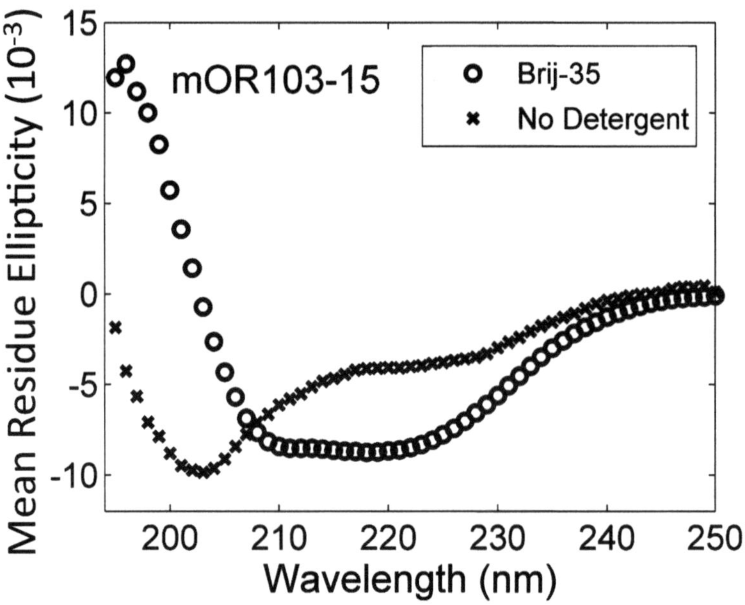

Fig. 2 Circular dichroism scan of mOR103-15 expressed in a cell-free reaction with and without detergent. The receptors expressed in a cell-free system could be purified and analyzed using circular dichroism. The scan showed a detergent is necessary for proper folding. Without a detergent, the purified protein has a random coil secondary structure. With a detergent (Brij-35), the secondary structure is helical, as expected for a GPCR (8)

Immunoaffinity chromatography and gel filtration chromatography can then be used to purify the expressed protein for structural and functional studies (Fig. 2).

2 Materials

2.1 Cell-Free Expression

1. *E. coli* lysate (RiNA GmbH, http://rina-gmbh.eu/, Qiagen and Life Technologies). The lysate should be stored at −80 °C, thawed on ice, and used within 4 h of defrosting. It can be refrozen and thawed once for optimal results. Aliquots can be made for larger volumes of lysate.
2. Reaction buffer. The buffer should be stored at −80 °C, thawed on ice, and used within 4 h of defrosting. It can be refrozen and thawed once for optimal results. Aliquots can be made for larger volumes of buffer. The specific buffer varies with each kit, but contains all essential amino acids, RNA polymerases, ribosomes, elongation factors, etc.
3. Sterile, DNase-free and RNase-free water.
4. Brij-35 (see Note 1).
5. Olfactory receptor gene ligated into the pIVex2.3 vector (Life Technologies) (see Note 2).

2.2 Rho1D4 Bead Coupling

1. Rho1D4 monoclonal antibody (Cell Essentials, http://www.cell-essentials.com/) at 2–8 mg/ml in coupling buffer. If the antibody is not in the coupling buffer upon arrival, it must first be dialyzed into the coupling buffer.
2. CNBr-activated Sepharose 4B beads (GE Healthcare).
3. Coupling buffer: 0.25 M $NaHCO_3$, 0.5 M NaCl, pH 8.3, in milliQ water.
4. HCl buffer: 1 mM HCl in milliQ water.
5. Ethanolamine buffer: 1 M ethanolamine, pH 8.0 in milliQ water.
6. Acetate buffer: 0.1 M NaOAc, 0.5 M NaCl, pH 4.0 in milliQ water.
7. Sodium azide buffer: 0.05 % NaN_3 in PBS, pH 7.2.

2.3 Receptor Purification

1. Rho1D4 monoclonal antibody-coupled sepharose beads.
2. DPBS (Life Technologies, 14190-250).
3. DNaseI (Life Technologies, 18047-019).
4. RNaseA (Life Technologies, 12091-039).
5. Sterile filtered water (0.22 μm) with a resistivity of at least 18 MΩ cm.
6. Wash buffer: 0.2 % fos-choline-14 (FC14) (Anatrace/Affymetrix). This is made from a 10 % FC14 stock solution in DPBS.

7. Elution buffer: 800 μM elution peptide Ac-TETSQVAPA-NH_2 (with an acetylated N-terminus and amidated C-terminus) dissolved in wash buffer.
8. High pH buffer: 0.1 M Tris–HCl, 0.5 M NaCl, pH 8.5.
9. Low pH buffer: 0.1 M sodium acetate, 0.5 M NaCl, pH 4.5.

2.4 Gel Filtration

1. Amicon Ultra Centrifugal Filter Units with Ultracel membranes (Millipore, 50 kDa MWCO membranes).
2. Sterile 96-well plates with V-shaped bottoms, ~500 μl/well capacity.
3. Wash buffer: 0.2 % FC14 in DPBS, sterile filtered through 0.22 μm filters.

3 Methods

Commercial cell-free expression kits can be used to produce milligrams of soluble, functional, and stable olfactory receptors. The most significant factor affecting expression levels and solubility is the choice of detergent. For example, the Brij family of detergents has repeatedly yielded high levels of receptors, while the fos-cholines, which are optimal for purification from HEK293 cells, result in almost no receptor expression (7, 8). This is likely due to their non-ionic nature, as ionic detergents can interfere with transcriptional and translational machinery. Factors like incubation temperature and time typically had negligible effects on expression. Although the optimal expression conditions must be validated for each protein of interest, and the solubility and functionality of the expressed receptors analyzed, the following protocol has proven to be optimal for most of the receptors that we have tried to express.

3.1 Cell-Free Protein Production

1. Thaw the *E. coli* lysate, reaction buffer, and DNA on ice.
2. Add 175 μl of the E. coli lysate to a sterile, DNAse-free, RNAse-free Eppendorf tube (see Note 3).
3. Add the plasmid to the lysate so that the final DNA concentration is 1 μg/100 μl.
4. Add DNAase-free and RNAse-free water so that the final volume of the cell-free reaction will be 500 μl.
5. Add Brij-35 to a final concentration of 0.2 % w/v (see Note 1).
6. Add 200 μl of the reaction buffer and mix thoroughly with a pipette.
7. Briefly spin down the eppendorf tube.
8. Place the cell-free reaction in an eppendorf rack in a shaking incubator at 33 °C and 250 rpm for 1 h (see Note 4).

9. After the reaction is complete, spin it down in a microcentrifuge for 5 min at 16,000×*g*.
10. Carefully transfer the supernatant to a fresh tube without disturbing the pellet. The supernatant contains solubilized receptor.
11. The synthesized receptors can be purified or run on an SDS-PAGE gel immediately, or stored at −20 °C for longer periods of time.

3.2 Rho1D4 Monoclonal Antibody-Sepharose Bead Coupling

1. Suspend the sepharose beads in HCl buffer. The hydrated beads will swell: 1 g of bead powder will yield approximately 3.5 ml of bead slurry.
2. Wash the beads for 15 min with the HCl buffer in a sintered glass funnel and vacuum flask. The beads should be resuspended in the buffer as they fall out of solution. After ~1 min, a vacuum should be applied until the liquid is removed. Do not overdry the beads. Approximately 200 ml of HCl buffer should be used per gram of bead powder. Several aliquots may be necessary.
3. Add the washed beads to the antibody in coupling buffer. Add 20 ml of bead slurry to 130–200 mg of antibody. The ratio should be 5–10 mg of antibody per ml of bead slurry.
4. Rotate the slurry/antibody solution at 4 °C until the antibody is bound to the beads. The antibody concentration in the supernatant can be monitored by measuring its absorbance at 280 nm. When the concentration is below 5 % of the original concentration, the binding reaction is complete. This procedure takes 4 h to overnight.
5. Remove the supernatant after the binding reaction is complete by spinning down the beads for 5 min at 800×*g*.
6. Remove excess antibody by washing the beads with 5 slurry volumes of coupling buffer.
7. Block remaining active groups with the ethanolamine buffer. Add a volume equal to the original supernatant volume and rotate overnight at 4 °C or 2 h at room temperature.
8. Remove excess uncoupled antibody by washing the beads four times, alternating between coupling buffer and acetate buffer. Use a sintered glass filter and a wash volume at least five times the original slurry volume.
9. Suspend the beads in 1 slurry volume of sodium azide buffer and store them at 4 °C.

3.3 Receptor Purification

1. Pipette the necessary amount of antibody-coupled beads into a fresh tube. Mix the beads first by gently shaking them to ensure that they are homogeneously suspended. The binding capacity

of fresh beads is ~0.7 mg/ml, and the capacity of regenerated beads is ~0.35 mg/ml.

2. Wash the beads with DPBS to remove excess sodium azide. Spin the beads down at 800×*g* for 1 min and then let them sit for a minute to allow them to completely settle to the bottom of the tube. Using a pipette, slowly remove the supernatant without disturbing the bead pellet. Add 1 bead volume of DPBS to pellet to resuspend the pellet. Repeat this process three times. After the last repetition, do not add more DPBS.
3. Add the supernatant from the cell-free reaction to the washed beads.
4. Add 1 μl of DNAse and 1 μl of RNAse for each ml of cell-free reaction volume.
5. Rotate the supernatant with the beads overnight at 4 °C to capture the synthesized protein.
6. After the overnight rotation, spin the beads at 800×*g* for 1 min and let them sit for 1 min to allow the bead pellet to settle. Remove the supernatant and transfer it to a tube labeled FT (Flow Through). Save a small sample of the FT for analysis and freeze the remainder at −80 °C in case the beads did not capture all of the synthesized receptors. Add 1 bead volume of wash buffer to the beads and rotate at 4 °C for 10 min.
7. Wash the OR-bound beads to remove any impurities. For each wash, spin the tube at 800×*g* for 1 min and allow it to sit for 1 min. Carefully remove the supernatant without disturbing the bead pellet and transfer it to a fresh tube (labeled Wash 1, Wash 2, etc.). Add 1 bead volume of wash buffer and rotate at 4 °C for 10 min. Repeat this process until the absorbance at 280 nm of the removed supernatant is less than 0.01 mg/ml. Typically, 13–20 washes are required. The washes can be run overnight at 4 °C if necessary.
8. Elute the synthesized ORs from the beads. Add 1 bead volume of elution buffer to the beads and rotate at room temperature for 1 h. Spin the beads at 800×*g* and let them sit for 1 min. Carefully remove the supernatant without disturbing the bead pellet and transfer it to a fresh, clean tube (labeled Elution 1, Elution 2, etc.). Repeat this process until the absorbance of the removed supernatant at 280 nm is less than 0.01 mg/ml. The supernatant contains the synthesized receptors. Typically, 5–7 elutions are required. Any of the elutions can be run overnight at 4 °C if necessary.
9. The washes and elutions can be stored at 4 °C until they are ready for use.
10. The elutions can be pooled and concentrated in centrifugal units with 50 kDa molecular weight cutoff filters. If residual

elution peptide must be removed (i.e., for circular dichroism), then the protein must be washed on the centrifugal units with excess wash buffer. 10 ml of wash buffer is usually sufficient to remove elution peptide from a concentrated protein sample with a total volume of 300 μl. If the receptors will be run on a size exclusion column, they must be concentrated to a volume that will fit in the loading loop. The receptors should be concentrated immediately prior to being loaded on the column to minimize aggregation and precipitation.

11. Used beads can be regenerated for reuse by washing them with 2–3 column volumes of alternating high pH and low pH buffers. This cycle should be repeated three times followed by re-equilibration in binding buffer.

3.4 Gel Filtration Chromatography

1. Equilibrate the gel filtration column with at least 1–2 column volumes of wash buffer. We use a HiLoad 16/60 Superdex 200 column (GE Healthcare) on an ÄKTA Purifier FPLC system (GE Healthcare).
2. Load the freshly concentrated OR sample into the system.
3. Run the system at 0.3 ml/min and monitor the UV absorbance at 215 and 280 nm. The monomeric form of our receptors typically exits the column at 60–65 ml. We collect the first 40 ml in a clean bottle. The remainder is collected in four 96-well V-bottom plates with 100 μl in each well.
4. Pool the appropriate eluted protein fractions together.
5. Concentrate the pooled fractions to the desired volume or concentration and store them at 4 °C until they are ready for further analysis. Samples can be kept at −80 °C for long-term storage and should only be thawed once as repeated freeze–thaw cycles can induce protein aggregation.

4 Notes

1. Brij-35 has been the optimal detergent in our experiments. However, other groups have found other detergents to be optimal for their GPCRs, especially other polyoxyethylenes related to Brij-35 (7). A preliminary detergent screen in which the cell-free reaction volumes are scaled down to a total volume of 25–50 μl may be necessary.
2. The OR genes have a 5′ *Nco*I site and a 3′ *Xho*I site for ligation. They also have a C-terminal rho1D4 epitope tag (TETSQVAPA) for purification followed by a double stop codon, and have potential glycosylation sites removed. The codons were optimized for *E. coli* expression.

3. The volumes listed here are for a total reaction volume of 500 μl. The volumes can be scaled up and down as necessary.
4. These are the optimal temperature, time, and rotation speed for the receptors we tested. We did not notice significant differences in OR yield with longer incubation times (up to 6 h), rotation speeds up to 300 rpm, and at temperatures between 30 and 37 °C. However, the optimal conditions may vary with different receptors. In particular, lower temperatures can increase the yield of soluble receptor, while higher temperatures can increase the total receptor yield.

References

1. Grisshammer R, Duckworth R, Henderson R (1993) Expression of a rat neurotensin receptor in *Escherichia coli*. Biochem J 295: 571–576
2. Hampe W et al (2000) Engineering of a proteolytically stable human β_2-adrenergic receptor/maltose-binding protein fusion and production of the chimeric protein in Escherichia coli and baculovirus infected insect cells. J Biotechnol 77:219–234
3. Sarramegna V, Talmont F, Demange P, Milon A (2003) Heterologous expression of G-protein-coupled receptors: comparison of expression systems from the standpoint of large-scale production and purification. Cell Mol Life Sci 60:1529–1546
4. McCusker EC, Bane SE, O'Malley MA, Robinson AS (2007) Heterologous GPCR expression: a bottleneck to obtaining crystal structures. Biotechnol Prog 23:540–547
5. Tate CG, Grisshammer R (1996) Heterologous expression of G-protein-coupled receptors. Trends Biotechnol 14:426–430
6. Ishihara G et al (2005) Expression of G protein coupled receptors in a cell-free translational system using detergents and thioredoxin-fusion vectors. Protein Expr Purif 41:27–37
7. Klammt C et al (2007) Cell-free production of G protein-coupled receptors for functional and structural studies. J Struct Biol 158:482–493
8. Corin K et al (2011) A robust and rapid method of producing soluble, stable, and functional G-protein coupled receptors. PLoS One 6(10): e23036
9. Kaiser L et al (2008) Efficient cell-free production of olfactory receptors: detergent optimization, structure, and ligand binding analyses. Proc Natl Acad Sci 105(41):15726–15731
10. Corin K et al (2011) Designer lipid-like peptides: a class of detergents for studying functional olfactory receptors using commercial cell-free systems. PLoS One 6(11):e25067
11. Wang X et al (2011) Peptide surfactants for cell-free production of functional G protein-coupled receptors. Proc Natl Acad Sci USA 108(22):9049–9054

Chapter 8

Human Olfactory Receptors: Recombinant Expression in the Baculovirus/Sf9 Insect Cell System, Functional Characterization, and Odorant Identification

Valéry Matarazzo and Catherine Ronin

Abstract

Cell surface expression of recombinant olfactory receptors (ORs) is a major limitation in characterizing their functional nature. We have shown that the recombinant expression of a human OR, OR 17-210, in the baculovirus/Sf9 insect cell system allows this protein to be expressed at the cell surface. We used Ca^{2+} imaging to demonstrate that recombinant OR 17-210 produces cellular activities upon odorant stimulation with ketones. Furthermore, this expression and functional system has been used to show that the preincubation of Human Odorant Binding Protein 2A decrease the calcium response of OR 17-210 following stimulation by acetophenone and beta ionone.

Key words Olfactory receptors, Odorants, Odorant binding protein, Baculovirus, Calcium imaging

1 Introduction

Mammalian olfactory receptors (ORs) are heterotrimeric guanine nucleotide-binding (G-protein)-coupled receptors localized at the cell-surface membrane at the cilia of olfactory sensory neurons. ORs recognize odorant molecules and translate the information present in chemical molecules into neural activity. To ensure early olfactory perception and discrimination of thousands of odorant molecules, ORs constitute the largest protein family among G protein-coupled receptors. They comprise approximately, 1,035 and 387 putatively functional ORs in mice and humans, respectively, also making it one of the largest gene families in the mammalian genome (1–3). Starting from gene sequences provided by genome analysis, the production and characterization of OR proteins represents a crucial step in assigning odorant ligands to specific ORs to elucidate olfactory encoding signals. However, odorant ligands are known for no more than 50 mammalian receptors. Difficulties in deciphering the functional nature of ORs have largely been the result of limited heterologous expression and impaired OR protein

Chiquito J. Crasto (ed.), *Olfactory Receptors: Methods and Protocols*, Methods in Molecular Biology, vol. 1003,
DOI 10.1007/978-1-62703-377-0_8, © Springer Science+Business Media, LLC 2013

translocation to the cell surface (4, 5). Also, OR specificity was found to be relatively broad: screening a practically infinite number of structurally diverse odorant molecules that carry information about the environment is quite challenging (6). To increase cell surface expression of OR binding sites and allow OR deorphanization and odorant ranking, we have developed a strategy based on the over expression of human OR protein in Sf9 cells together with Gα protein in order to transduce a measurable biological signal in response to odorant stimulation. Quantitative pharmacological odorant screening for ORs has been performed using an in vivo Ca^{2+} imaging methodology. The human Odorant Binding Protein 2A (hOBP-2A) is expressed in the nasal mucus and thought to transport odorants through the nasal mucus to ORs (7). This protein can bind diverse types of odorants such as acetophenone and beta ionone with a higher affinity. The Ca^{2+} imaging methodology applied here to investigate the function of hOBP-2A on activation of the human OR 17-210 by these two odorants.

2 Materials

2.1 Recombinant Baculovirus

1. TC 100 medium containing 5 % of Fetal Bovine Serum (FBS). FBS must be inactivated in a hot water bath at 56 °C for 30 min.
2. DOTAP reagent (Roche).
3. Low melting agarose is mixed extemporaneously with TC 100 medium.
4. Patented recombinant baculovirus (WO 2004/011658) and modified pGmAc217T transfer vector were obtained from Prof. Gérard DEVAUCHELLE: Professor of Virology and Molecular Virology at the Universities of Montpellier and Nîmes (France) and Dr. Martine DUONOR-CERRUTI: Research Director at the National Scientific Research Center (CNRS). Dr Martine Cerruti is the head of the “Baculovirus and Therapy laboratory” of CNRS at Saint-Christol-lès Alès (France).

2.2 Immunofluorescence and Confocal Microscopy

1. Round microscope coverslips (12 mm diameter).
2. Phosphate buffered saline (PBS): Prepare 10× stock with 1.37 M NaCl, 27 mM KCl, 100 mM Na_2HPO_4, 18 mM KH_2PO_4 (adjust to pH 7.4 with HCl if necessary) and autoclave before storage at room temperature. Prepare a working solution by diluting with water to 1:9 ratio.
3. Paraformaldehyde: Prepare a 4 % (w/v) solution in PBS fresh for each experiment. The solution has to be carefully heated (use a stirring hot plate in a fume hood) for complete dissolution, and then cooled to room temperature for use.

4. Quench solution: 50 mM NH_4Cl in PBS.
5. Permeabilization solution: 0.2 % (v/v) Triton X-100 in PBS.
6. Primary antibody dilution buffer: 4 % goat serum in PBS.
7. Monoclonal mouse anti-FLAG and FITC-conjugated anti-mouse (Sigma) (see Note 4).
8. Rhodamine-conjugated anti-HA (Roche).
9. Mounting medium: Mowiol (Calbiochem): Add 2.4 g of Mowiol with 6 g of Glycerol and 6 ml of distilled sterile water. Stir thoroughly and leave 2 h at room temperature. Add 12 ml of 0.2 M Tris–HCl pH 8.5 and incubate at 53 °C until the Mowiol has dissolved. For fluorescence, add 1,4-diazobicyclo-(2.2.2)-octane (DABCO) to 2.5 % to reduce fading. Clarify by centrifugation at 5,000×*g* for 20 min. Transfer supernatant in aliquots of 1 ml and store at −20 °C. Once defrosted, the solution is stable for at least 1 month.

2.3 Measurement of Intracellular Calcium

1. Fura 2-AM (Invitrogen) is stored at −20 °C and extemporarily prepared at 4 μM.
2. MBS medium (2-(*N*-morpholino) ethanesulfonic acid-buffered saline) contains 10 mM NaCl, 60 mM KCl, 25 mM $MgCl_2$, 1.8 mM $CaCl_2$, 4 mM D-Glucose, 110 mM sucrose, and 10 mM 2-(*N*-morpholino) ethanesulfonic acid. The pH must be adjusted to 6.2 at room temperature with trizma base.
3. Open diamond bath chamber for round coverslips (RC-25) and its corresponding platform is adapted for 12 mm diameter round coverslip (Warner instruments). Bath dimension is: 25×12×1.1 mm.
4. Polyethylene tubing (PE-160, Warner instruments) is connected to the chamber for perfusion and suction.
5. The calcium imaging Olympus microscope (Germany) is equipped with multiway wavelength illumination system POLYCHROME II T.I.L.L Photonics (GmbH, Planegg, Germany) for excitation. Images are acquired with one long-working distance lens (×20). Spatiotemporal Ca^{2+}-distributions are investigated using a PCO interline camera. The wavelength for the excitation of the dye is alternated between 340 and 380 nm and is adapted to the exposure time of the camera (20–40 ms; binning: 2). Acquisition and calculation of the fluorescence images are done using WinNT based-software (T.I.L.L-photonics).

2.4 Odorant Screening

1. All odorants were obtained from Henkel (Düsseldorf, Germany) and diluted at 50 μM in MBS medium as a starting concentration (see Note 2).
2. Minimal set composition: camphor, limonene, isoamyl acetate, acetophenone, 2-heptanone, and anisole.

3. Lactones set composition: pentadecanolide, whisky lactone.
4. Ketones set composition: acetophenone, camphor, β-ionone.
5. Esters set composition: benzyl acetate, 8-cineole, isobornyl acetate, hexyl acetate, benzyl benzoate, isoamyl acetate, ethyl butyrate, 2-phenyl ethyl acetate.
6. Phenols set composition: thymol, eugenol, guaiacol.
7. Alcohols set composition: menthol, citronellol, geraniol, linalool, octanol, benzyl alcohol, cinnamyl alcohol, phenyl ethanol, 1-octanol.
8. Aldehydes set composition: citral, amyl cinnamaldehyde, cinnamaldehyde, piperonal.
9. Positive control for cell survival activity: Octopamine (Sigma) is prepared at 50 μM in MBS.

2.5 Preincubation with Recombinant hOBP-2A and Activation of OR 17-210

1. Recombinant hOBP-2A (7) produced in *Pichia pastoris* was a kind gift from Dr. J.C. Pernollet (NOPA, INRA, Jouy-en-Josas). It is stored in 50 mM PBS at −20 °C in presence of 2 % glycerol.
2. hOBP-2A was tested from 100 up to 1,000 nM in MBS medium.
3. Highly purified acetophenone and beta ionone were a kind gift from Dr. Thierry Thomas-Danguin (FLAVIC INRA, Dijon). These odorants were dissolved in 1 mM DMSO (Sigma-Aldrich) and further diluted at 100 μM in MBS medium.

3 Methods

The insect cell line Sf9 (ATCCCRL 1711), which is derived from the ovary of the fall armyworm *Spodoptera Frugiperda* (8) can be infected by reconstructed baculovirus *Autographa californica* (Ac) nuclear polyhedrosis virus. RCPGs can be functionally expressed in the plasma membrane of Sf9 cells (9, 10). We initially used this approach to express human ORs with G_{16} protein.

Since translocation of OR to the cell surface is a critical step during recombinant expression, we made an oligonucleotide containing a signal peptide and a FLAG tag in order to facilitate translocation of OR to the cell surface and assess translocation efficiency. Thus, an 108 pb oligonucleotide, corresponding to the EGT (ecdysteroid UDPS-glycosyltransferase) signal peptide (11) from AcMNPV followed by the FLAG epitope tagging sequence and two unique restriction sites (NcoI and KpnI), were inserted into the BglII/SacI sites of the pGmAc217T transfer vector (12). This transfer vector named pGmAc217T-FLAG-OR allows insertion of any OR into the NcoI and KpnI sites.

The G-protein produced has also been tagged with the HA epitope, and thus, verification of OR insertion at the plasma membrane and co-localization with the G-protein can be accomplished through confocal indirect immunofluorescence. OR activity can be validated by imaging calcium variations of the cell in response to chemical odorant sets.

3.1 Recombinant Baculovirus

We previously constructed a recombinant baculovirus, designed as AcSLP10Gα16Ph, expressing Gα16 and the polyhedrin proteins which for the last produce an occlusion-bodies positive (ob^+) virus phenotype (13). In order to obtain double-recombinant baculovirus, AcSLP10Gα16-Ph DNA is co-transfected with pGmAc217T-FLAG-OR transfer vector into Sf9 cells. The final double recombinant baculovirus is designated as AcGα16–OR and produce an occlusion-bodies negative (ob^-) virus phenotype (Fig. 1).

1. Sf9 insect cells are cultured on flasks and maintained at 27 °C in a modified TC 100 medium containing 5 % of FBS. They are subcultured two times weekly.
2. Cells are passed 3 days before transfection (see Note 1).
3. Prepare viral DNA from AcSLP10Gα16-Ph from viral particles as previously described (14).
4. Prepare DNA from pGmAc217T-FLAG-OR transfer vector.
5. Transfections are initiated using 500 ng of baculoviruses DNA and 10 μg of transfer vector. Viral and plasmid DNAs are mixed with 40 μl of DOTAP in 3 ml of TC 100 medium without serum and added to 2×10^6 cells.
6. After 4 h of incubation at 28 °C, the transfection mixed is removed from the cells and replaced with 5 ml of TC 100 medium supplemented with 5 % FBS.
7. The recombinant baculovirus are collected from the supernatant of transfected cells after 5 days of incubation at 28 °C.
8. Infect new Sf9 cells by adding 1 ml of the recombinant baculovirus.
9. After 1 h, recover cells by a 2 % agarose-TC 100 medium supplemented with 5 % FBS. This forming gel medium will limit virus propagation allowing virus purification.
10. After 3 days of infection, collect the occlusion-bodies negative (ob^-) virus phenotype using a Pasteur pipette. This virus phenotype corresponds to the final double recombinant baculovirus designated as AcGα16–OR.
11. Repeat step 10 twice in order to purify the baculovirus.
12. Amplify virus by successive cell infections. Recombinant baculoviruses can be stored at 4 °C until use.

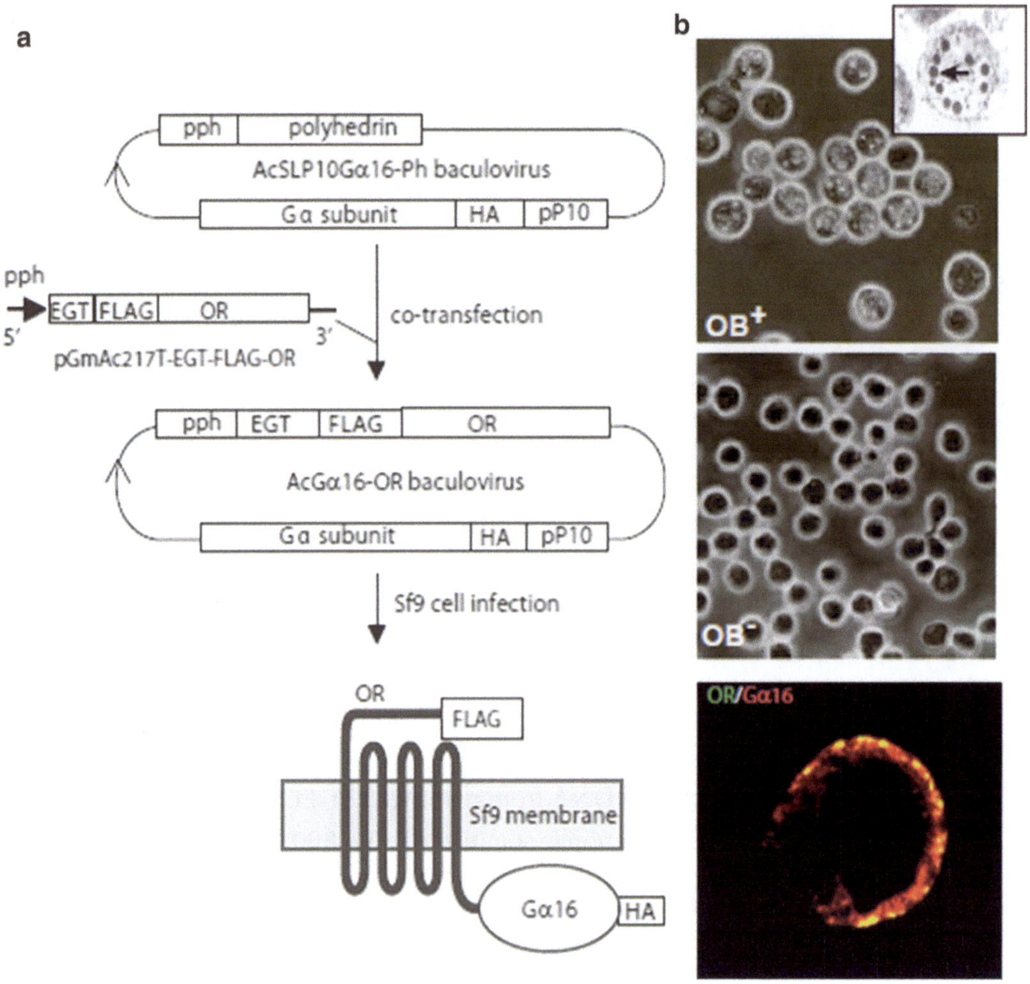

Fig. 1 (**a**) Schematic representation for the construction of a double recombinant virus. (**b**) Image of a contrast phase acquisition of Sf9 cells showing occlusion-bodies positive (ob^{+}) after infection with AcSLP10Gα16Ph virus (*top panel*) and occlusion-bodies negative (ob^{-}) after infection with AcSLP10Gα16-OR double recombinant virus (*middle panel*). *Bottom panel*: Confocal microscopy images of Sf9 cells infected with the AcSLP10Gα16-OR baculovirus. Cells co-expressing OR and Gα16 are visualized using an anti-FLAG mAb (green fluorescence) and an anti-HA-rhodamine-conjugated mAb (red fluorescence) respectively

3.2 Immuno-fluorescence and Confocal Microscopy

1. Prepare 12 mm sterilized glass coverslips by holding with tweezers, addition of 95 % ethanol, and passing through the flame of a Bunsen burner and then placed in the culture dishes to cool.
2. Sf9 cells are cultured on sterilized glass coverslips at 27 °C in TC 100 medium supplemented with 5 % FBS for 48 h at 27 °C.
3. 1 ml of AcGα16–OR viral solution (5 pfu/cellule) is added and cells are kept at 27 °C for 36 h at 27 °C.

4. All fixation steps must be conducted on ice to prevent receptor internalization.
5. Post-infected cells are washed twice with PBS and fixed with 4 % of paraformaldehyde for 20 min.
6. The paraformaldehyde is discarded (into a hazardous waste container) and the samples are washed twice for 5 min each with PBS.
7. Residual paraformaldehyde is quenched by incubation in 0.05 M NH4Cl-PBS for 10 min at room temperature followed by an additional two washes with PBS.
8. Cells are blocked with 10 % inactivated goat serum buffer for 30 min.
9. Cells are incubated for 1 h at room temperature with a mouse anti-FLAG (1:500) then washed twice times for 10 min with PBS.
10. A secondary antibody FITC-conjugated anti-mouse (1:128) is incubated for 1 h at room temperature then washed twice times for 10 min with PBS.
11. Cells are fixed again with 3 % of paraformaldehyde and permeabilized with 0.1 % of Triton X-100 for 10 min.
12. Cells are blocked again with 10 % inactivated goat serum buffer for 30 min and incubated for 1 h with rhodamine-conjugated anti-HA (1:20).
13. The samples are washed twice times for 5 min with PBS and then aspirated dry from one corner.
14. Coverslips are dropped into sterile water to remove salt.
15. The samples are then ready to be mounted. The slip of the coverslip is carefully inverted into a drop of mounting anti-fading Mowiol medium on a microscope slide (see Note 3). The samples are air-dried for 24 h and placed at 4 °C for confocal analysis.
16. The slides are visualized under phase contrast microscopy (to locate the cells and identify the focal plane) and under confocal microscopy with an oil-×63 lens. Excitation at 488 nm induces the FITC fluorescence (green emission) for the FLAG-OR, while excitation at 465 nm induces rhodamine fluorescence (red emission) for the HA-Gα16. Examples of the immunostaining are shown in Fig. 1.

3.3 Measurement of Intracellular Calcium Concentration Under Resting State and Following OR Activation

1. 12 mm sterilized glass coverslips are prepared by holding with tweezers, adding 95 % ethanol, and passing through the flame of a Bunsen burner and then placed in the culture dishes to cool.
2. Sf9 cells are cultured on sterilized glass coverslips at 27 °C in TC 100 medium supplemented with 5 % inactivated FBS for 48 h at 27 °C.

3. 1 ml of AcGα16–OR viral solution (5 pfu/cellule) is added and the cells are kept at 27 °C for 36 h. at 27 °C. Noninfected cells must be prepared for control experiments.
4. The culture medium is removed and washed with MBS medium.
5. The supernatant is removed and the cells are incubated for 1 h at dark-room temperature with a MBS solution containing Fura 2-AM.
6. The cell cultures are washed with an excess of Fura 2-AM free MBS and mounted in a bath chamber (Fig. 2).
7. The bath chamber is placed on the stage of the microscope and the culture is gravity-perfused with MBS medium at 0.85 ml/min (3.78 cm/min) and at room temperature (23 °C).
8. Dual-wavelength (340/380 nm) measurements of Fura 2-AM fluorescence are performed using a calcium imaging Olympus microscope (Germany). Example of the Fura 2-AM fluorescence excited at 340 and 380 nm are shown in Fig. 2.
9. The Fluorescence ratio (f_{340}/f_{380}) is calculated to determine Ca^{2+} intracellular variations using Till-Photonics software (see Notes 5 and 7).

3.4 Odorant Screening

1. Prepare odorants to be tested within hours in 50 μM in MBS medium and keep them on ice until use.
2. Fill four distinct syringes with 10 ml of MBS, 2 ml of the minimal set solutions, 2 ml of chemically defined groups and 2 ml octopamine, respectively.
3. Solutions are applied to the bath by means of a manual switch-tap, ensuring that each syringe are switched off and started by switching on the MBS syringe (see Note 8).
4. Dual-wavelength (340/380 nm) measurements are begun while cells are bathed with MBS medium (see Note 9).
5. Odotope screening using successive sets of odorants is started. The MBS medium syringe settings are switched off and the minimal set syringe is switched on.
6. A calcium response is generally observed less than 1 min after odorants incubation and characterized by an increase of the f_{340}/f_{380} ratio for cell expressing OR-210 while control cells (uninfected or infected with a wild type baculovirus) do not respond.
7. After 1 min of exposure, the syringe switched off to stop odorant stimulation.
8. Cells are rinsed off with 1 ml of MBS medium by switching-on the MBS syringe.
9. When calcium imaging responses go back to the resting state, another stimulus can be applied by repeating steps 5–9.

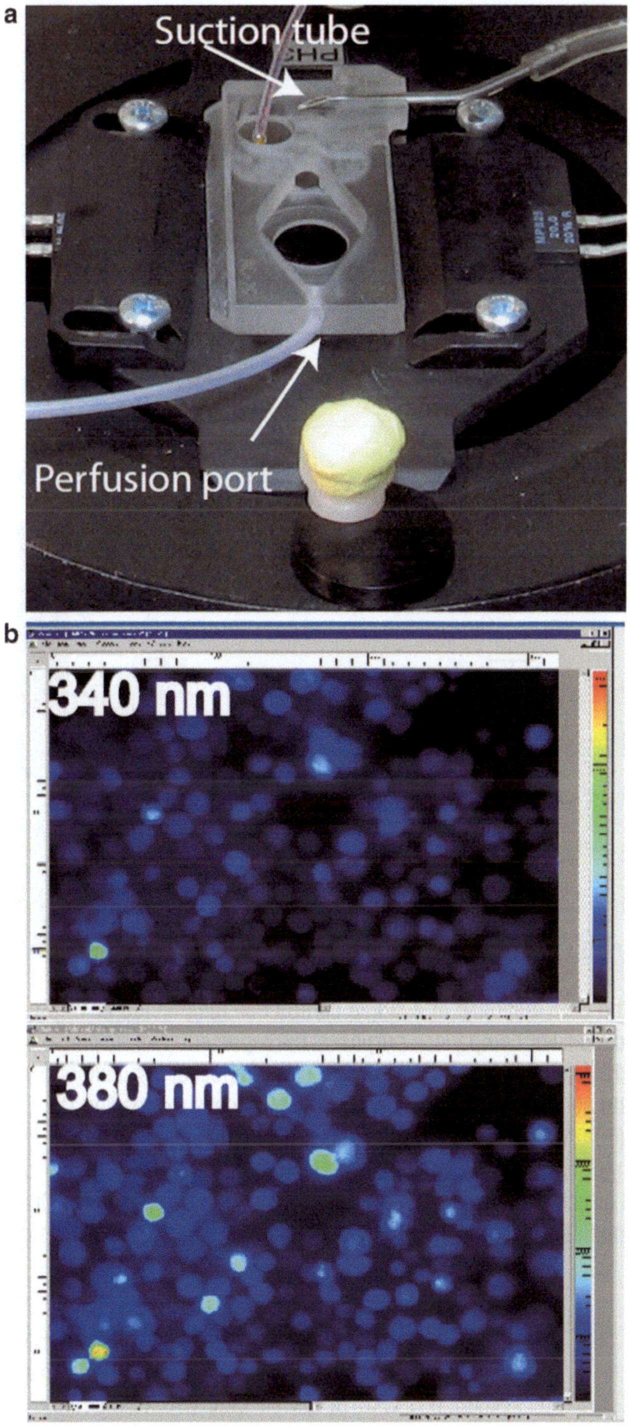

Fig. 2 (**a**) Open diamond bath chamber for round coverslips (RC-25, Warner instrument) and its corresponding platform. The chamber is placed to the stage of an inverted microscope. 12 mm round glass coverslip is sealed to the center of the chamber using tech grease. (**b**) Image captures of Sf9 cells loaded with Fura 2-AM and excited at 340 and 380 nm. (**c**) Calcium response of Sf9 cells after minimal set stimulation. Typical recording

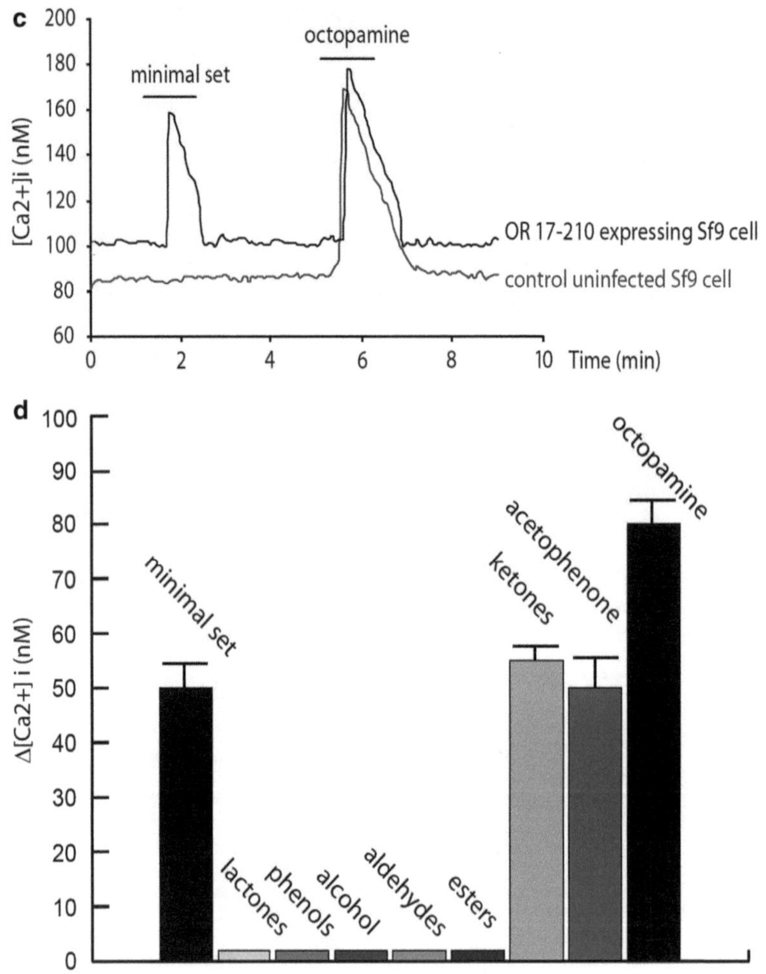

Fig. 2 (continued) of the Ca^{2+} response of a single control cell (uninfected or wild type baculovirus infected) after minimal set application. For comparison purposes, the representative response of a control cell with a cell infected with OR 17-210 recombinant baculovirus are superimposed. In all experiments, cell responsiveness is assessed by a second stimulation with octopamine. (**d**) Sf9 cells expressing OR 17-210 are stimulated by different odorant sets (50 μM) or individual molecules (50 μM) as indicated on the *top* of each *bar graph*. For each Sf9 cell, the value of $\Delta(Ca^{2+})i$ is calculated. It represents the value of the maximal $(Ca^{2+})i$ increase obtained upon odorant or octopamine application (50 μM) minus the value of $(Ca^{2+})i$ at the resting state. Data are expressed as the mean ± SEM. Student's *t*-test is used to determine statistical significance

10. At the end of each assay, apply 50 μM of octopamine to the chamber in order to stimulate the endogenous GPCR (9) to control the intactness of infected or uninfected Sf9 cells (see Note 6). Examples of calcium responses elicited by the odorant and octopamine are shown in Fig. 2 .
11. All responding cells in the field of the camera are recorded and at least 100 individual cells by such experiment have been considered for statistical analysis. Intracellular calcium concentration has been averaged from four typical experiments.

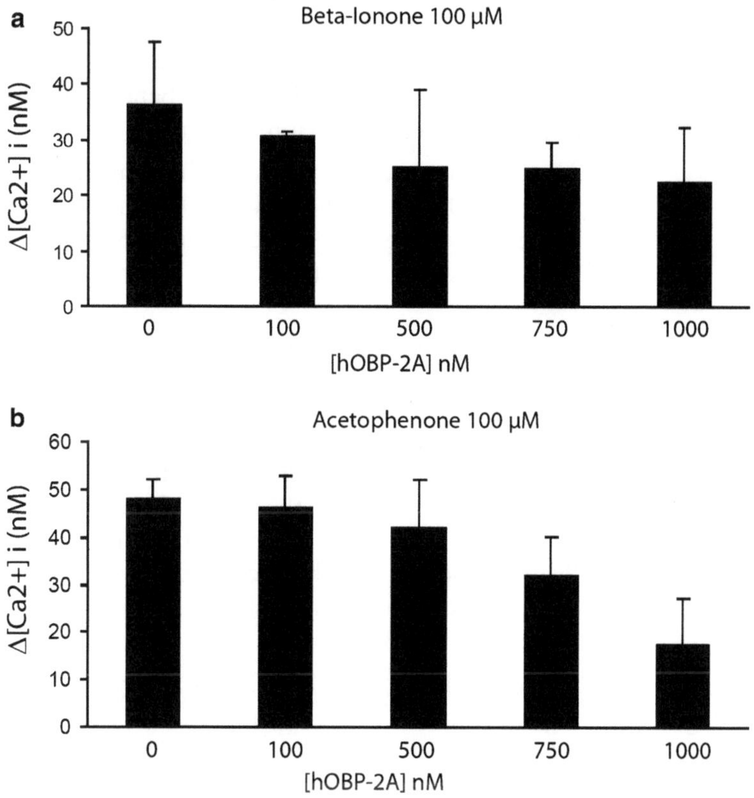

Fig. 3 Effect of hOBP-2A on OR 17-210 activation by 100 μM beta ionone (**a**) and acetophenone (**b**). The data are expressed as the mean ± SEM of calcium response on an average of 100 positive cells individually recorded

3.5 Preincubation with Recombinant hOBP-2A and OR Activation

1. Sf9 cells are cultured on sterilized glass coverslips at 27 °C in TC 100 medium supplemented with 5 % inactivated FBS for 48 h at 27 °C.
2. 1 ml of AcGα16–OR viral solution (5 pfu/cellule) is added and the cells are kept at 27 °C for 24 h at 27 °C. Noninfected cells must be prepared for control experiments.
3. Before experiments, the culture medium is removed and washed with MBS medium.
4. The supernatant is removed and the cells are incubated 1 h at darkroom temperature with a MBS solution containing Fura 2-AM.
5. The cell cultures are washed with an excess of Fura 2-AM free MBS and mounted in a bath chamber.
6. hOBP-2A is incubated with OR 17-210 expressing cells (Fig. 3). Dose response curves are established by using a single glass coverslip by OBP concentration. Increasing concentration of 100, 500, 750 and 1,000 nM are routinely tested.

Each dose was applied in quadruplicate (four coverslips). Controls are carried out by omitting OBP (see Note 10).

7. The best agonists of OR 17-210, beta ionone and acetophenone, are tested for calcium response.
8. All responding cells in the field of the camera are recorded and at least 100 individual cells by such experiment are considered for statistical analysis. Intracellular calcium concentration has been averaged from 4 to 6 typical experiments.

4 Notes

1. Since Sf9 cells are loosely attached, trypsinization is not required. Using a cell scraper, gently scrape the cells off the bottom of the flask into the media. Check that all the cells have come off by inspecting the base of the flask before moving on. From a confluent flask of cells, perform a 1:3 split ratio to obtain a new confluent flask of cells in 3 days. Avoid doing more than 1:10 split ratio.
2. Be aware that manipulating odorants generates a significant, unpleasant smell in the laboratory. We recommend manipulating odorants under a laboratory fume hood.
3. Air bubbles are undesirable when mounting coverslips on a microscope slide. A slow and careful pressure application on the coverslip can be applied to remove them.
4. We have found these antibodies to generate virtually no artifactual background in immunofluorescence. Numerous similar reagents are available from other commercial sources.
5. This calcium imaging protocol can be adapted for many other cell culture systems.
6. Calcium recording can be assessed during 30 min to 1 h. After that time, new cell cultures on coverslip, loaded with Fura 2-AM should be used. We recommend preparing cells loaded with Fura 2-AM every 20 min.
7. $(Ca^{2+})i$ is calculated from the formula proposed by Grynkiewicz et al. (15). Minimal and maximal experimental ratio values (Rmin and Rmax) are calculated for Sf9 cells. Rmax and Rmin are the values of R under saturating (5 μM A23187, a Ca^{2+}-ionophore) and Ca^{2+}-free conditions (4 mM EGTA combined with 5 μM A23187), respectively. The $(Ca^{2+})i$ of individual cells is thus calculated every 5 s and values are plotted versus time. Resting $(Ca^{2+})i$ values of individual Sf9 cells oscillated between 80 and 100 nM. For comparison, we determined the ΔCa^{2+} for each cell, defined as the value of the $(Ca^{2+})i$ increase after ligand application minus the value of $(Ca^{2+})i$ at the resting

state. For each experiment, the most responsive cells (5–10 cells out of 20–30 in the camera field of view) are selected.

8. Air bubbles can form when perfusing solutions through the imaging chamber. They should be removed before the experiment is started.
9. No difference in basal calcium concentration should be observed between noninfected cells, infected cells preincubated or not with the OBP.
10. Recombinant hOBP-2 is sensitive to artifactual proteolysis and is not a stable protein. We recommend keeping it frozen in very small aliquots so as to thaw 1–2 aliquots by calcium imaging experiments.

Acknowledgments

The authors would like to thank Pr. Gérard DEVAUCHELLE (University of Montpellier, France), Dr Martine CERRUTI (CNRS, France) for virology experiments. We are also indebted to Dr. Thierry Thomas-Danguin and Dr Anne Tromlin (FLAVIC INRA, Dijon) for fruitful discussion on odorant mixtures and odor perception in humans. The authors are grateful to Dr Olivier Clot-Faybesse and Dr Brice Marcet for calcium imaging screening and Maryse Gremigni, Aurélie Seassau, and Nadia Elfangoud for experiments with hOBP-2. The work has been supported by a grant from a PNRA program of the French National Agency (ANR AROMALIM, coordinated by R. Gervais).

References

1. Malnic B, Godfrey PA, Buck LB (2004) The human olfactory receptor gene family. Proc Natl Acad Sci U S A 101(8):2584–2589
2. Saito H et al (2009) Odor coding by a mammalian receptor repertoire. Sci Signal 2(60):ra9
3. Zhang X, Firestein S (2002) The olfactory receptor gene superfamily of the mouse. Nat Neurosci 5(2):124–133
4. Bush CF, Hall RA (2008) Olfactory receptor trafficking to the plasma membrane. Cell Mol Life Sci 65(15):2289–2295
5. McClintock TS, Sammeta N (2003) Trafficking prerogatives of olfactory receptors. Neuroreport 14(12):1547–1552
6. Malnic B (2007) Searching for the ligands of odorant receptors. Mol Neurobiol 35(2): 175–181
7. Tcatchoff L et al (2006) A single lysyl residue defines the binding specificity of a human odorant-binding protein for aldehydes. FEBS Lett 580(8):2102–2108
8. Vaughn JL et al (1977) The establishment of two cell lines from the insect Spodoptera frugiperda (Lepidoptera; Noctuidae). In Vitro 13(4):213–217
9. Hu Y, Rajan L, Schilling WP (1994) Ca2+ signaling in Sf9 insect cells and the functional expression of a rat brain M5 muscarinic receptor. Am J Physiol 266(6 Pt 1):C1736–C1743
10. Raming K et al (1993) Cloning and expression of odorant receptors. Nature 361(6410):353–356
11. Riegel CI, Lanner-Herrera C, Slavicek JM (1994) Identification and characterization of the ecdysteroid UDP-glucosyltransferase gene

of the Lymantria dispar multinucleocapsid nuclear polyhedrosis virus. J Gen Virol 75(Pt 4):829–838

12. Gaymard F et al (1996) The baculovirus/insect cell system as an alternative to Xenopus oocytes. First characterization of the AKT1 K+channel from Arabidopsis thaliana. J Biol Chem 271(37):22863–22870
13. Matarazzo V et al (2005) Functional characterization of two human olfactory receptors expressed in the baculovirus Sf9 insect cell system. Chem Senses 30(3):195–207
14. Poul MA et al (1995) Cassette baculovirus vectors for the production of chimeric, humanized, or human antibodies in insect cells. Eur J Immunol 25(7):2005–2009
15. Grynkiewicz G, Poenie M, Tsien RY (1985) A new generation of Ca2+ indicators with greatly improved fluorescence properties. J Biol Chem 260(6):3440–3450

Chapter 9

Single Cell RT-PCR Identification of Odorant Receptors Expressed by Olfactory Neurons

Bettina Malnic

Abstract

Mammals have between 400 and 1,300 functional odorant receptor (OR) genes in their genomes. Each olfactory sensory neuron in the nose expresses only one single type of OR out of this vast repertoire. The OR expressed by an olfactory sensory neuron determines its functional activity and wiring to the olfactory bulb. Therefore, the identification of the OR type expressed by individual neurons contributes to the understanding of the mechanisms of odorant perception. Here we describe the protocol that combines single cell RT-PCR and degenerate PCR to identify the OR expressed by single olfactory neurons. In a primary PCR reaction, cDNAs corresponding to all mRNAs present in a single neuron are synthesized. In a secondary PCR reaction, the cDNAs corresponding to the OR expressed by this neuron are amplified using degenerate primers that match OR sequences.

Key words Odorant receptors, Olfactory neurons, Single cell RT-PCR, Degenerate PCR, Olfactory epithelium

1 Introduction

Odorant receptors (ORs) are encoded by large gene families whose sizes vary among different species (1). The mouse olfactory gene family, for example, is composed of over 1,000 intact OR genes and 328 OR pseudogenes (1, 2). It has been demonstrated via a series of different approaches, including *in situ* hybridization (3, 4), limiting dilution and RT-PCR (5), single cell RT-PCR (6), and transgenic mice (7), that each olfactory sensory neuron expresses only one single type of OR. Consequently, the receptor type that is expressed determines the range of odorants to which the neuron will respond. The axons of olfactory neurons that express one same type of OR converge into one or few glomeruli at two specific sites of the olfactory bulb (8–10). ORs also play important roles in the projection of the olfactory neuron axons to specific glomeruli (11, 12).

Chiquito J. Crasto (ed.), *Olfactory Receptors: Methods and Protocols*, Methods in Molecular Biology, vol. 1003,
DOI 10.1007/978-1-62703-377-0_9,

In order to understand how an odorant is perceived it is necessary to determine the specificities of individual ORs. Different methods have been used to link ORs to the odorants they recognize (13). One of these consists in combining calcium imaging and single cell RT-PCR to determine the ORs expressed by olfactory neurons that respond to different odorant ligands (6, 14): Olfactory sensory neurons are dissociated and loaded with calcium sensitive dyes like fura-2, and sequentially exposed to a panel of different odorants. Once the odorants in the panel that activate one given olfactory neuron are registered, the responsive neuron is transferred with the help of a microcapillary pipette to a microcentrifuge tube and the OR expressed by this neuron is determined using single cell RT-PCR.

Even though many olfactory genes (such as OMP, Gαolf, and ACIII) are expressed by mature olfactory sensory neurons, some genes are differentially expressed by different olfactory neurons. Olfactory neurons located in the zones 2–4 of the epithelium, for example, express the olfactory cell adhesion molecule (OCAM), while the neurons located in the apical zone 1 do not (15). Other genes have been shown to be differentially expressed in different zones in the olfactory epithelium (16–20). Some olfactory neurons express different types of the trace amine associated receptors (TAARS), instead of ORs (21). Single cell cDNA libraries can be prepared from different olfactory sensory neurons in order to identify genes that are differentially expressed by these neurons. Genes that are expressed in different developmental stages (progenitors, immature or mature olfactory neurons) can also be identified using this same approach (16, 17).

Here we describe the protocol that combines single cell RT-PCR (based on refs. 6, 22–24) and a secondary PCR reaction with degenerate primers to identify the OR expressed by single olfactory neurons (illustrated in Fig. 1).

2 Materials

1. Trypsin-EDTA, Invitrogen cat. # R-001-100.
2. M-MLV (Moloney Murine Leukemia Virus) Reverse Transcriptase, Invitrogen, cat. # 28025013.
3. AMV (Avian Myeloblastosis Virus) Reverse Transcriptase, Invitrogen, cat. #12328019.
4. Terminal deoxynucleotidyl transferase, Invitrogen, cat. # 10533065.
5. dNTP set (100 mM each), GE Healthcare, cat. # 28-4065-51.
6. BSA (albumin), 20 mg/mL, Roche, cat. # 10711454001.

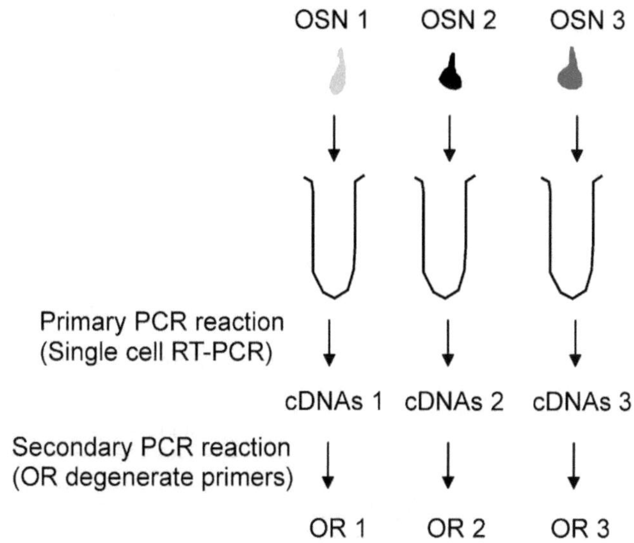

Fig. 1 Schematic representation of the steps followed to identify the OR expressed by one single olfactory neuron. Single neurons are transferred to microtubes and subjected to a single cell RT-PCR reaction (primary PCR reaction). The cDNAs amplified in this step are next used as template in a secondary PCR reaction with degenerate primers that match to conserved sequences in ORs, to identify the OR that is expressed in the given olfactory neuron

7. UltraPure DNAse–RNAse-free distilled water, Invitrogen, cat. # 10977023.
8. AmpliTaq DNA polymerase LD (Low DNA), Applied Biosystems, cat. # N8080158.
9. (dT)-X amplification oligonucleotide primer AL1, Oligos, etc.: 5′ATTGGATCCAGGCCGCTCTGGACAAAATATGAATTC (T)24 3′.
10. Nonidet P40 (NP-40), Roche cat. # 11754599001.
11. RNAguard Ribonuclease Inhibitor (Human placenta), GE Healthcare. Prime RNase Inhibitor, Utech Products, Inc., cat. # 2143457.
12. Oligomer d(T)24, Oligos, etc.
13. Triton X-100 (Sigma).
14. Taq DNA polymerase, 5 U/μL, Roche, cat. # 11146165001.

3 Methods

3.1 Preparation of Single Cells

1. Dissociate small pieces of mouse olfactory epithelium tissue for 10 min at 37 °C in 2 mL trypsin–EDTA (0.025 % trypsin and 0.01 % EDTA in PBS pH 7.4).

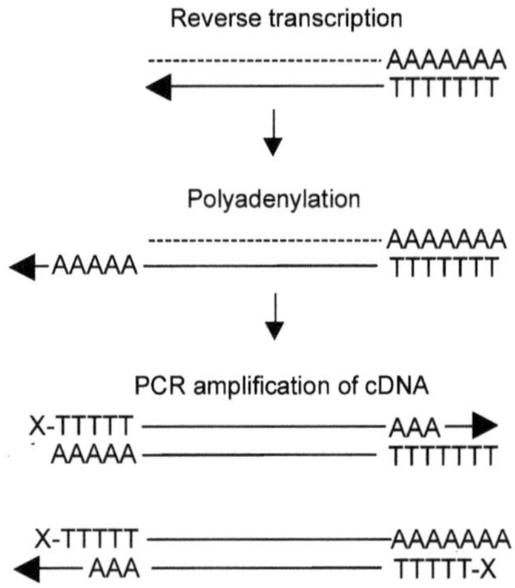

Fig. 2 Single cell RT-PCR (primary PCR reaction). cDNAs corresponding to all mRNAs in a single cell are synthesized in three steps: reverse transcription, polyadenylation, and PCR amplification

2. Add 1 mL Dulbecco's Modified Eagle Medium (DMEM) plus 10 % heat inactivated fetal bovine serum (FBS). Gently pipette up and down several times with the help of a heat-polished Pasteur pipette.
3. Collect cells by centrifugation and resuspend in 2 mL of PBS containing 0.1 % BSA. Place cells in a small petri dish and observe under an inverted microscope.

3.2 Single Cell RT-PCR (Primary PCR Reaction)

The single cell RT-PCR procedure is divided into three basic steps: reverse transcription, tailing, and PCR (Fig. 2, see Notes 2 and 3).

Reverse transcription

1. Pick single cells under a microscope with the help of a pulled microcapillary pipette (50 μL, Fisher) and transfer them to a microcentrifuge tube containing 4 μL of ice-cold cell lysis mix (see solutions bellow, see Note 1).
2. Incubate the tube at 65 °C for 1 min to promote unfolding of the mRNA.
3. Cool the tube to room temperature for 3 min to allow annealing of the oligomer d(T)24 primer. Place on ice.
4. Add 0.5 μL of a mixture containing 50 U of M-MLV and 1 U of AMV reverse transcriptases and incubate for 30 min at 37 °C.
5. Stop the reaction by heating to 65 °C for 10 min. Cool on ice for 2 min.

Tailing

6. Add an equal volume (5 μL) of 2× tailing buffer (see below) containing 10 U of terminal deoxynucleotidyl transferase.

7. Incubate for 15 min at 37 °C.
8. Stop the reaction by heating to 65 °C for 10 min.

PCR

9. Add the following PCR mix to a final volume of 100 μL:

   (for one sample):

   10 μL 10× AmpliTaq DNA polymerase LD buffer (supplied with the enzyme).

   10 μL 25 mM $MgCl_2$ (supplied with the enzyme).

   1 μL of each 100 mM dNTPs (4 μL total).

   0.5 μL 2 % BSA.

   0.05 μL Triton X-100.

   1 μL (dT)-X amplification oligonucleotide primer AL1 (5 μg/μL).

   62.5 μL DNAse–RNAse-free distilled water.

   2 μL (10 U) AmpliTaq DNA polymerase LD.

10. Subject the reaction to the following PCR cycles:

    One cycle of 1.5 min at 96 °C.

    25 cycles of 1 min at 96 °C, 2 min at 42 °C, and 6 min at 72 °C with 10 s extensions per cycle.

    One cycle of 10 min at 72 °C.

    Soak at 4 °C.

11. Add 5 U of AmpliTaq DNA Polymerase LD and subject the reaction to the additional following cycles:

    One cycle of 1.5 min at 96 °C.

    20 cycles of 1 min at 96 °C, 1 min at 42 °C, and 2 min at 72 °C with 10 s extensions per cycle.

    One cycle of 10 min at 72 °C.

    Soak at 4 °C.

12. Store the samples at −70 °C.

    Solutions used in the primary PCR reaction:

    Cell lysis mix: is composed of 96 μL of cell lysis buffer*; 2 μL RNAse inhibitors (1 μL of RNAguard and 1 μL of Prime RNAse inhibitor) and 2 μL of cDNA primer mix# freshly diluted 1:24 in DNAse–RNAse-free distilled water.

    *Cell lysis buffer: 208 μL 5× M-MLV buffer; 5.2 μL NP40; 786.8 μL DNAse–RNAse-free distilled water.

    #cDNA primer mix: 12.5 mM each dNTP; 6.125 OD_{260}/mL oligo d(T)24.

    2× *Tailing buffer*: 400 μL 5× terminal transferase buffer (supplied with the enzyme); 15 μL 100 mM dATP; 585 μL DNAse–RNAse-free distilled water.

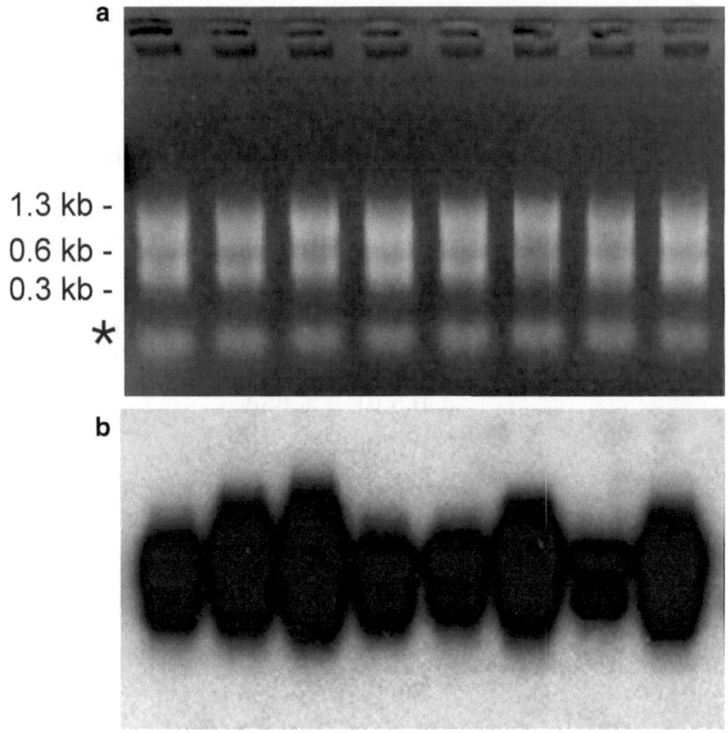

Fig. 3 Analysis of the primary PCR products. (**a**) 1.5 % agarose gel showing the amplified cDNAs, which range from 0.5 to 1 kb in length. The *lower band* (*) corresponds to residual AL1 primers. (**b**) The agarose gel shown in (**a**) was blotted and the blot hybridized to a probe prepared from OMP, which is abundantly and specifically expressed in mature olfactory neurons

3.3 Analysis of the Single Cell RT-PCR Products

Successful amplification of the single cell RT-PCR products can be analyzed by electrophoresis on a 1.5 % agarose gel. 5 μL of each sample should be enough to visualize the cDNA products, which should range from 0.5 to 1 kb in length. The agarose gel in Fig. 3a shows the single cell RT-PCR products obtained from eight different olfactory neurons. Genes expressed in each sample can be analyzed by conventional Southern blotting. Figure 3b shows a Southern blot of the samples shown in (A) hybridized to a probe for Olfactory Marker Protein (OMP), a specific marker for mature olfactory neurons.

3.4 Secondary PCR Reaction

In order to identify the OR expressed in each olfactory neuron, a secondary PCR reaction using degenerate primers that match OR sequences is performed according to the following steps:

1. Add 1 μL of a 1/100 dilution of the primary PCR reaction to the following PCR mix (final volume of 50 μL):

   5 μL 10× Taq DNA polymerase buffer with $MgCl_2$ (final 1.5 mM $MgCl_2$).

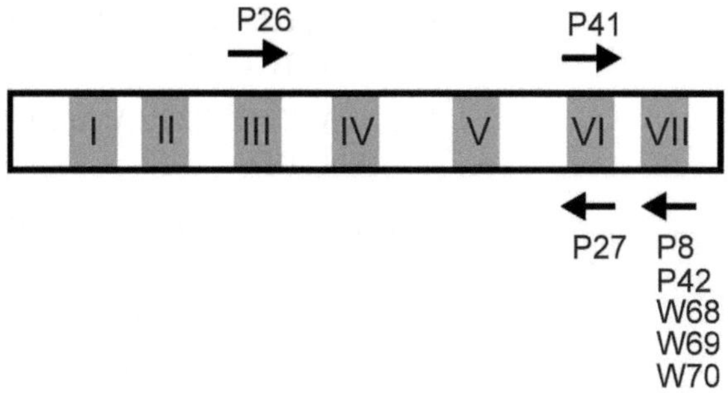

Fig. 4 Degenerate primers used in the secondary PCR reaction. The OR transmembrane (TM) regions matched by the different degenerate primers are indicated

1.0 μL 10 mM dNTP mix (final 0.2 mM each dNTP).

2.0 μL P26 degenerate primer (stock 50 μM, final 2 μM).

2.0 μL P27 degenerate primer (stock 50 μM, final 2 μM).

0.5 μL Taq DNA polymerase (Roche, 2.5 U).

37.0 μL dH_2O.

2. Subject the reaction to the following PCR cycles:

One cycle of 1.5 min at 96 °C.

45 cycles of 1 min at 96 °C, 3 min at 40 °C, and 3 min at 72 °C.

One cycle of 10 min at 72 °C.

Soak at 4 °C.

Different sets of degenerate primers can be used to amplify the OR cDNAs. The P26 and P27 primers (6, 25) match respectively conserved regions in TM3 and TM6 (Fig. 4). The region between TM6 and TM7 can be amplified by using the P41 (matching TM6) and P42, W68, W69, W70, or P8 (matching TM7) primers (6) (Fig. 4). Also, because the complete repertoire of mouse ORs is now available (2, 26–28), additional degenerate primers that match sequences not covered by the primers above, could be designed (see Notes 4 and 5).

The sequences of the primers are:

P26: GCITA(C/T)GA(C/T)CGITA(C/T)GTIGCIATITG

P27: ACIACIGAIAG(G/A)TGIGAI(G/C)C(G/A)CAIGT

P41: AA(GA)(TG)CITTI(AGT)(AC)IACITG(CT)G(GC)ITCICA

P42: TC(TC)(TC)TIGTI(TC)TI(AG)(TC)IC(TC)GATAIAT IATIGG(GA)TT

W68:TCI(TC)T(GA)TTIC(TG)IAGIG(TA)(GA)TAIAT(GA) AAIGG(GA)TT

W69: TC(TC)TT(GA)TTIC(TG)IAGIG(TA)(GA)TAIA(TC) IA(GC)IGG(GA)TT

W70: TCIT(GC)(GA)TTIC(TG)IA(GA)I(GC)A(GA)TAIA TIATIGG(GA)TT

P8: (GA)TTIC(TG)IA(AG)I(GC)(TA)(GA)TAIAT(AG)AAIGG (GA)TT

Anchor sequences (such as T7 and T3) can be added to the 5′ ends of the primers to allow direct sequencing of the PCR products. PCR products are gel purified and sequenced using an anchor sequence primer and the ABI PRISM Big dye terminator cycle sequencing kit. PCR products amplified using the TM3/TM6 primers should be around 390 bp long and PCR products amplified using the TM6/TM7 primers should be around 180 bp.

4 Notes

1. For the experiment aims to be successfully achieved, it is important to be sure that only one single olfactory neuron is picked. To confirm this, the cell can be transferred to a fresh wash buffer (PBS plus 01 % BSA), picked again, and transferred to the microcentrifuge tube containing the lysis buffer.
2. During the primary and secondary PCR reactions it is extremely important to avoid contamination, including contamination of one sample with other samples. Every precaution to avoid contamination in PCR reactions should be strictly followed. These include: the use of gloves, the use of aerosol barrier micropipette tips, the use of clean DNA free bench and accessories (tube racks, ice buckets, etc.). Microcentrifuge tubes should be kept closed as much as possible. Stock solutions should be stored as small aliquots at −70 °C.
3. The entire procedure should be performed as quickly as possible, without interruptions, in order to avoid RNA degradation.
4. For the samples where no PCR product is obtained in the secondary reaction when using the TM3/TM6 primers, different combinations of the TM6/TM7 primers could be tried. Since the primary PCR reaction amplifies cDNAs prepared from the 3′ ends of mRNAs, the use of primers that match regions located towards the 3′ end could facilitate amplification of OR cDNAs with long 3′ untranslated regions. If a PCR product is still not obtained when using the TM6/TM7 primers, it is possible that the OR expressed in this olfactory neuron has a sequence not well matched by the primers described here. Additional primers could be designed, using the information from all known OR sequences. Also, it is possible that this OR mRNA has a very long 3′ end, so that its coding region is not represented in the cDNA pool. In this case, no primer will work.

5. Finally, some of the olfactory neurons do not express OR genes, but other types of receptors, such as TAARs (21), and guanylyl cyclase D (29, 30). In this case, no PCR products will be obtained when using OR degenerate primers.

References

1. Nei M, Niimura Y, Nozawa M (2008) The evolution of animal chemosensory receptor gene repertoires: roles of chance and necessity. Nat Rev Genet 9:951–963
2. Godfrey PA, Malnic B, Buck LB (2004) The mouse olfactory receptor gene family. Proc Natl Acad Sci U S A 101:2156–2161
3. Ressler KJ, Sullivan SL, Buck LB (1993) A zonal organization of odorant receptor gene expression in the olfactory epithelium. Cell 73:597–609
4. Vassar R, Ngai J, Axel R (1993) Spatial segregation of odorant receptor expression in the mammalian olfactory epithelium. Cell 74:309–318
5. Chess A, Simon I, Cedar H, Axel R (1994) Allelic inactivation regulates olfactory receptor gene expression. Cell 78:823–834
6. Malnic B, Hirono J, Sato T, Buck LB (1999) Combinatorial receptor codes for odors. Cell 96:713–723
7. Ishii T, Shou S, Kohda A, Nakatani H, Shiroishi T, Okumura K, Iwakura Y, Nagawa F, Tsuboi A, Sakano H (2001) Monoallelic expression of the odorant receptor gene and axonal projection of olfactory sensory neurons. Genes Cells 6:71–78
8. Mombaerts P, Wang F, Dulac C, Chao S, Nemes A, Mendelsohn M, Edmondson J, Axel R (1996) Visualizing an olfactory sensory map. Cell 87:675–686
9. Ressler KJ, Sullivan SL, Buck LB (1994) Information coding in the olfactory system: evidence for a stereotyped and highly organized epitope map in the olfactory bulb. Cell 79:1245–1255
10. Vassar R, Chao S, Sitcheran R, Nunez J, Vosshall L, Axel R (1994) Topographic organization of sensory projections to the olfactory bulb. Cell 79:981–991
11. Mombaerts P (2006) Axonal wiring in the mouse olfactory system. Annu Rev Cell Dev Biol 22:713–737
12. Wang F, Nemes A, Mendelsohn M, Axel R (1998) Odorant receptors govern the formation of a precise topographic map. Cell 93:47–60
13. Malnic B (2007) Searching for the ligands of odorant receptors. Mol Neurobiol 35:175–181
14. Touhara K, Sengoku S, Inaki K, Tsuboi A, Hirono J, Sato T, Sakano H, Haga T (1999) Functional identification and reconstitution of an odorant receptor in single olfactory neurons. Proc Natl Acad Sci U S A 96:4040–4045
15. Yoshihara Y, Kawasaki M, Tamada A, Fujita H, Hayashi H, Kagamiyami H, Mori K (1997) OCAM: a new member of the neural cell adhesion molecule family related to zone-to-zone projection of olfactory and vomeronasal axons. J Neurosci 17:5830–5842
16. Tietjen I, Rihel J, Cao Y, Koentges G, Zakhary L, Dulac C (2003) Single-cell transcriptional analysis of neuronal progenitors. Neuron 38:161–175
17. Yu T, McIntyre J, Bose S, Hardin D, Owen M, McClintock T (2005) Differentially expressed transcripts from phenotypically identified olfactory sensory neurons. J Comp Neurol 483:251–261
18. Norlin EM, Alenius M, Gussing F, Hagglund M, Vedin V, Bohm S (2001) Evidence for gradients of gene expression correlating with zonal topography of the olfactory sensory map. Mol Cell Neurosci 18:283–295
19. Norlin M, Berghard A (2001) Spatially restricted expression of regulators of G-protein signaling in primary olfactory neurons. Mol Cell Neurosci 17:872–882
20. Gussing F, Bohm S (2004) NQO1 activity in the main and the accessory olfactory systems correlates with the zonal topography of projection maps. Eur J Neurosci 19:2511–2518
21. Liberles S, Buck L (2006) A second class of chemosensory receptors in the olfactory epithelium. Nature 442:645–650
22. Brady G, Iscove N (1993) Construction of cDNA libraries from single cells. Methods Enzymol 225:611–621
23. Dulac C, Axel R (1995) A novel family of genes encoding putative pheromone receptors in mammals. Cell 83:195–206
24. Matsunami H, Buck L (1997) A multigene family encoding a diverse array of putative pheromone receptors in mammals. Cell 90:775–784
25. Michaloski J, Galante P, Malnic B (2006) Identification of potential regulatory motifs in odorant receptor genes by analysis of promoter sequences. Genome Res 16:1091–1098

26. Niimura Y, Nei M (2005) Evolutionary dynamics of olfactory receptor genes in fishes and tetrapods. Proc Natl Acad Sci U S A 102:6039–6044
27. Young JM, Friedman C, Williams EM, Ross JA, Tonnes-Priddy L, Trask BJ (2002) Different evolutionary processes shaped the mouse and human olfactory receptor gene families. Hum Mol Genet 11:535–546
28. Zhang X, Rodriguez I, Mombaerts P, Firestein S (2004) Odorant and vomeronasal receptor genes in two mouse genome assemblies. Genomics 83:802–811
29. Fulle HJ, Vassar R, Foster DC, Yang RB, Axel R, Garbers DL (1995) A receptor guanylyl cyclase expressed specifically in olfactory sensory neurons. Proc Natl Acad Sci U S A 92:3571–3575
30. Juilfs DM, Fulle HJ, Zhao AZ, Houslay MD, Garbers DL, Beavo JA (1997) A subset of olfactory neurons that selectively express cGMP-stimulated phosphodiesterase (PDE2) and guanylyl cyclase-D define a unique olfactory signal transduction pathway. Proc Natl Acad Sci U S A 94:3388–3395

Chapter 10

Next-Generation Sequencing of the Human Olfactory Receptors

Joel D. Mainland, Jason R. Willer, Hiroaki Matsunami, and Nicholas Katsanis

Abstract

Humans have approximately 400 intact olfactory receptors (ORs). Among this set there are a large number of variations between individuals, a subset of which affects receptor function and can lead to interindividual variation in olfactory perception. Technological progress and cost erosion in next-generation sequencing have given us the opportunity to determine the sequence of the entire OR gene set with high fidelity and to measure the extent of variation in this functional module across many individuals. Given that whole genome sequencing remains prohibitively expensive for this purpose, especially since the OR sub-genome represents only ~0.0125 % of the human genome, we have designed a targeted capture method to enrich the OR for next-generation sequencing, which we describe here. Using this method we have been able to sequence an individual's OR sub-genome with high coverage, enabling us to identify variation with high sensitivity and specificity. This method can be used to accurate assess the amount of variability in this module and to identify the functional role of individual ORs in olfactory perception.

Key words Illumina, Next-generation sequencing, Olfactory receptor, Olfaction, Odor, SNP, Targeted genomic enrichment

1 Introduction

Linking olfactory receptor (OR) responses to perceptual phenotypes has proven challenging. In animal models where ORs can be modified genetically, the olfactory percept can only be queried indirectly. Due to the overlap in response properties for olfactory neurons, electrophysiological techniques have difficulty teasing apart which OR leads to a given neuron's response. Although knockout animals or GFP-labeled knock-in animals have been helpful, those techniques are suited for looking at one or a few ORs at a time due to the time and resources involved in creating each animal. Here we take advantage of both next-generation sequencing technologies and hybrid selection capture to develop a

Chiquito J. Crasto (ed.), *Olfactory Receptors: Methods and Protocols*, Methods in Molecular Biology, vol. 1003,
DOI 10.1007/978-1-62703-377-0_10, © Springer Science+Business Media, LLC 2013

method for examining the relationship between genetic variation in ORs and perceptual variation.

Until recently, sequencing the entire olfactory sub-genome for a large subject cohort was prohibitively expensive. Here, we present a targeted selection protocol which, together with emerging next-generation sequencing technology, can be used to generate a comprehensive catalog of the sequence variation of olfactory sub-genome and to relate OR gene variation to olfactory perception. By examining the relationship between OR and perception at this scale, these methods have the potential to uncover fundamental principles underlying olfactory perception.

2 Materials

2.1 Library Preparation

1. RNA-free and protein-free genomic DNA sample, 2 μg.
2. TE buffer.
3. Covaris shearing tubes.
4. PCR Plate Adhesive Seal.
5. TruSeq™ DNA Sample Prep Kit (Illumina).
6. UltraPure™ DNase–RNase-Free Distilled Water.
7. RNase–DNase-free PCR plates.
8. Ethanol 96 %.
9. Picogreen kit (Invitrogen).
10. Solid phase reversible immobilization (SPRI) paramagnetic beads (Beckman Coulter) (see Note 1).
11. TAE buffer.
12. DNA Ladder.
13. Agarose.
14. RediLoad loading buffer (Invitrogen).
15. Ethidium bromide (CAUTION: Mutagen and potential carcinogen).
16. Magnetic stand.
17. Phusion 2× Master Mix (Thermo Fisher).

    Primers 1 and 2

    Primer 1: 5′ AATGATACGGCGACCACCGAGA 3′

    Primer 2: 5′ CAAGCAGAAGACGGCATACGAG 3′

2.2 Targeted Enrichment

1. SureSelect Target Enrichment System for Illumina Paired-End Sequencing Library (Agilent).
2. Picogreen kit (Invitrogen).
3. Sealable tubes.

4. Nutator.
5. Vortex mixer.
6. Incubator.
7. Dynal MyOne Streptavidin T1 magnetic beads (Invitrogen).
8. SPRI paramagnetic beads.
9. LoBind 1.5 mL tube.

2.3 Sequencing

1. TruSeq PE Cluster Kit (Illumina PE-203-5001 or PE-300-2001).
2. TruSeq SBS Kit v5 (Illumina FC-104-5001 ×4).
3. GAIIx; Recipe_GA2-PEM2X_MP_75 + 7 + 75Cycle_v8.3.

2.4 Sequencing Analysis

You will need to install the following free software packages:

1. BWA (http://bio-bwa.sourceforge.net/).
2. GATK (http://www.broadinstitute.org/gatk/download), with resource bundle.
3. Picard (http://picard.sourceforge.net).
4. Samtools (http://samtools.sourceforge.net/).

3 Methods

3.1 Library Preparation (See Note 2)

Experimental design (sequence coverage estimation): (see Fig. 1 for overview).

An Illumina GA IIx outputs at least 320 million mapped reads per run (1). About 85 % of those will map to the genome. A 150 bp paired end design means we will end up with about 96 Gb of reads per run. About 40 % of the reads (40 Gb) will map to the target region. For a target size of 4,00,000 bp, the coverage for one sample would be ~8,500×. The consensus in the field is that 15× coverage is necessary to call heterozygous individuals with high confidence. Note that setting your mean coverage to 15× will result in large portions of your target region being covered at less than 15×. Our typical goal is to have 95 % of the target nucleotides covered at 15×. Pilot work indicates that to achieve this goal we need to set our mean coverage to 300×. It is therefore possible to multiplex up to 280 subjects per flow cell, or 35 per lane.

3.1.1 Scaling the Protocol

This protocol can be scaled up or down to fit different cohorts. For example, for a cohort of ten subjects (or multiples of 10), a 100 bp paired-end run on one lane of a GA IIx would result in 300× coverage. In the other direction, since the development of this protocol Illumina has released both the HiSeq 2000 sequencer and a commercial kit with 96 barcodes. Multiplexing 96 subjects in each

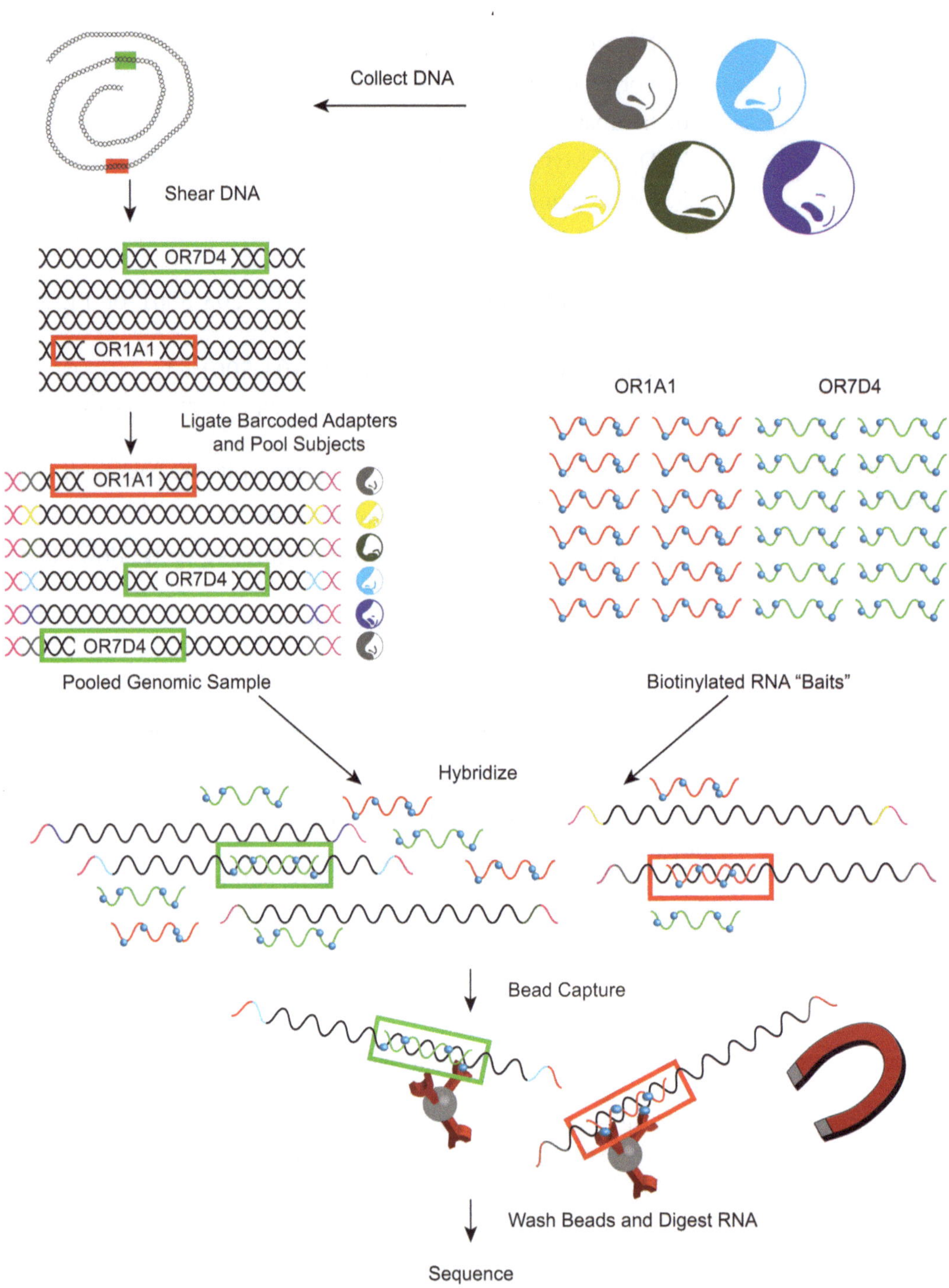

Fig. 1 The figure illustrates an overview of the protocol. DNA from each subject is collected and sheared. Barcoded adapters are ligated to the ends for downstream processing and sample disambiguation. Biotinylated RNA molecules matching odorant receptor genes are hybridized with the genomic DNA. Pieces of DNA with sequences complementary to the RNA baits are pulled down using magnetic streptavidin-coated beads. The unbound DNA is washed away, the RNA is digested, and the captured DNA is eluted and ready for downstream sequencing

lane using a 100 bp paired end run would give a projected mean coverage of 410× per subject. Note that these scaled-down and scaled-up protocols have not yet been tested. We recommend against reads shorter than 75 bp due to the sequence similarity across OR genes. We also recommend caution with the commercial 96 barcode kit, as the longer adapters may interfere with capture efficiency (2).

3.1.2 Shear DNA

Measure concentration of genomic DNA using Picogreen kit (see Note 3), following manufacturer's instructions.

1. Dilute 3 μg of DNA in TE Buffer to a total volume of 120 μL.
2. Following manufacturer's instructions, shear DNA on a Covaris using the following settings:

Duty cycle	10 %
Intensity	5
Bursts/s	200
Cycles	6 of 60 s each
Mode	Frequency sweeping

3. Centrigue the Covaris tube to 600 × *g* for 5 s.
4. Transfer 50 μL of fragmented DNA to each well of a new PCR plate.

3.1.3 Perform End Repair

1. Add 40 μL of End Repair Mix (included in the TruSeq kit) to each well containing fragmented DNA. Pipette up and down gently to mix thoroughly. Change the tip after each sample.
2. Seal the plate with an adhesive seal. Incubate the plate on a thermal cycler for 30 min at 30 °C.

3.1.4 Purify Sample

1. Add 160 μL of well-mixed SPRI Beads to each well of the plate.
2. Pipette up and down to mix thoroughly. Change the tip after each sample.
3. Incubate the plate at room temperature for 15 min.
4. Place the plate on the magnetic stand at room temperature for 15 min.
5. Remove 127.5 μL of supernatant, taking care not to disturb the beads. Change the tip after each sample.
6. Repeat step 5 once. Some liquid may remain in each well.
7. With the plate on the magnetic stand, add 200 μL of freshly prepared 80 % EtOH to each well without disturbing the beads.

8. Incubate the plate at room temperature for 30 s, then remove and discard all supernatant. Take care not to disturb the beads. Change the tip after each sample.
9. Repeat steps 7 and 8 for a total of two 80 % EtOH washes.
10. Remove plate from magnetic stand and let the plate dry at room temperature for 15 min.
11. Resuspend the dry pellet in 17.5 μL resuspension buffer. Gently pipette up and down ten times to mix thoroughly.
12. Incubate at room temperature for 2 min.
13. Place the plate on the magnetic stand at room temperature for 5 min or until liquid appears clear.
14. Transfer 15 μL of the clear supernatant from each well of the plate to the corresponding well of a new PCR plate (see Note 4).

3.1.5 Adenylate 3′ Ends

1. Add 12.5 μL of A-Tailing Mix and 2.5 μL of resuspension buffer (both included in the TruSeq kit) to each well.
2. Mix up and down using pipette, changing tip after each sample.
3. Apply adhesive seal and incubate on thermal cycler for 30 min at 37 °C.

3.1.6 Ligate Adapters

1. Add 2.5 μL of DNA Ligase mix (included in the TruSeq kit) to each well of the plate.
2. Add 2.5 μL of each thawed DNA Adapter Index (included in the TruSeq kit) to the appropriate wells.
3. Mix up and down using pipette, changing tip after each sample.
4. Apply adhesive seal and incubate for 10 min at 30 °C.
5. Remove plate from thermal cycler and add 5 μL of Stop Ligase Mix (included in the TruSeq kit).
6. Mix up and down using pipette, changing tip after each sample.
7. Add 42.5 μL of well-mixed SPRI Beads to each well of the plate.
8. Mix up and down using pipette, changing tip after each sample.
9. Incubate at room temperature for 15 min.
10. Place the plate on the magnetic stand at room temperature for 2 min.
11. Remove and discard 80 μL of supernatant from each well. Change the tip after each sample.
12. Add 200 μL of freshly prepared 80 % EtOH to each well without disturbing the beads.

13. Incubate the plate at room temperature for at least 30 s, then remove and discard all supernatant.
14. Repeat steps 12 and 13 once for a total of two 80 % EtOH washes.
15. Let the plate dry at room temperature for 15 min.
16. Remove from magnetic stand, then resuspend with 52.5 μL Resuspension Buffer (included in the TruSeq kit).
17. Mix up and down using pipette, changing tip after each sample.
18. Incubate the plate at room temperature for 2 min.
19. Place the plate on the magnetic stand at room temperature for 2 min.
20. Transfer 50 μL of the clear supernatant to the corresponding well of a new PCR plate. Change the tip after each sample.
21. Add 50 μL of well-mixed SPRI Beads to each well for a second clean up.
22. Mix up and down using pipette, changing tip after each sample.
23. Incubate the plate at room temperature for 15 min.
24. Place the plate on the magnetic stand at room temperature for 2 min.
25. Remove and discard 95 μL of supernatant from each well. Some liquid may remain in each well. Change the tip after each sample.
26. With the plate remaining on the magnetic stand, add 200 μL of freshly prepared 80 % EtOH to each well without disturbing the beads.
27. Incubate the plate at room temperature for at least 30 s, then remove and discard all supernatant. Change the tip after each sample.
28. Repeat steps 26 and 27 once for a total of two 80 % EtOH washes.
29. Let the plate dry at room temperature for 15 min, then remove from the magnetic stand.
30. Resuspend with 22.5 μL Resuspension Buffer (included in the TruSeq kit).
31. Mix up and down using pipette, changing tip after each sample.
32. Incubate at room temperature for 2 min.
33. Place the plate on the magnetic stand at room temperature for 2 min.
34. Transfer 20 μL of the clear supernatant to the corresponding well in a new PCR plate, changing the tip after each sample.

3.1.7 PCR Amplification

1. Add 5 μL of thawed PCR Primer Cocktail (included in the TruSeq kit) to each well of the PCR plate. Change the tip after each sample.
2. Add 25 μL of thawed PCR Master Mix (included in the TruSeq kit) to each well of the PCR plate. Mix up and down using pipette, changing tip after each sample.
3. Amplify the plate in a thermal cycler using the following protocol:

   13 cycles of:

   (a) 98 °C for 10 s.

   (b) 60 °C for 30 s.

   (c) 72 °C for 30 s.

   72 °C for 10 s.

   Hold at 4 °C.
4. Run samples in a 2 % agarose gel at 150 V for 30 min to check amplification levels. If a smear around 200–400 bp is visible, continue with step 5. Otherwise, place the sample back into the thermocycler and run the following protocol:

   Three cycles of:

   (a) 98 °C for 10 s.

   (b) 60 °C for 30 s.

   (c) 72 °C for 30 s.

   72 °C for 10 s.

   Hold at 4 °C.
5. Add 50 μL of well-mixed SPRI Beads to each well of the PCR plate. Mix up and down using pipette, changing tip after each sample.
6. Incubate the plate at room temperature for 15 min.
7. Place the plate on the magnetic stand at room temperature for 2 min.
8. Remove and discard 95 μL of supernatant from each well; some liquid may remain in each well. Change the tip after each sample.
9. With the plate remaining on the magnetic stand, add 200 μL of freshly prepared 80 % EtOH to each well without disturbing the beads.
10. Incubate the plate at room temperature for at least 30 s, then remove and discard all supernatant. Change the tip after each sample.
11. Repeat previous two steps once for a total of two 80 % EtOH washes.

12. Let the plate dry at room temperature for 15 min, then remove from the magnetic stand.
13. Resuspend with 32.5 μL Resuspension Buffer (included in the TruSeq kit).
14. Mix up and down using pipette, changing tip after each sample.
15. Incubate at room temperature for 2 min.
16. Place the plate on the magnetic stand at room temperature for 2 min.
17. Transfer 30 μL of the clear supernatant to the corresponding well in a new PCR plate.
18. Change the tip after each sample.

3.1.8 Pool Subjects According to Experimental Design

1. Capture reagents are much more cost effective when barcoded samples are pooled prior to capture.
2. Quantify DNA concentration using Picogreen or another method of equivalent accuracy
3. You will need 500 ng total DNA per lane in 3.4 μL of nuclease-free water. If samples are too dilute, dry diluted samples using vacuum centrifuge and resuspend in nuclease-free water.

3.2 Targeted Enrichment

3.2.1 Hybridization

There are a number of solutions for targeted enrichment (3). Here we use solution-based enrichment (4) in the form of the Agilent SureSelect commercial kit. Probes are designed using the Agilent eArray software (see Note 5). We used design ELID #0352781 (see Note 6).

1. Quantify DNA concentration using Picogreen.
2. Add 3.4 μL of pooled library at 147 ng/μL to an Eppendorf tube.
3. Make Buffer from the SureSelect kit reagents, volume for one capture reaction

Hyb 1	25 μL
Hyb 2	1 μL
Hyb 3	10 μL
Hyb 4	13 μL

   Add to 8-strip A and incubate at 65 °C.

4. Make Bait on ice, volume for one capture reaction

Sure Select library	2 μL
RNase Block	0.5 μL
Nuclease-free water	4.5 μL

5. Make Block Mix from the SureSelect kit reagents, volume for one capture reaction

#1	2.5 μL
#2	2.5 μL
#3	0.6 μL

6. Add 3.4 μL of Library to 8-strip B.
7. Add 5.6 μL of Block Mix to 8-strip B.
8. Mix, seal plate.
9. Incubate at 95 °C for 5 min.
10. Keep plate at 65 °C for the rest of the protocol.
11. Add 40 μL hybridization buffer to each well of 8-strip A.
    Wait for 5 min.
12. Add 7 μL Bait to each well of 8-strip C
    Seal, wait 2 min.
13. Add 13 μL of A (Buffer) to C (Bait).
14. Add 9 μL (all) of B (Library) to C (Bait) and mix well.
15. Seal and incubate 72 h.

3.2.2 Purify Post-Capture

1. Resuspend Dynal MyOne Streptavidin T1 (Invitrogen) magnetic beads on a vortex mixer.
2. Add 50 μL Dynal magnetic beads to a 1.5 mL tube for each capture.
3. Wash beads three times:
    (a) Add 200 μL of binding buffer.
    (b) Vortex, place on magnetic stand.
    (c) Remove and discard supernatant.
4. Estimate and record the volume that remained after incubation.
5. Add the hybridization mix directly from the thermal cycler to the bead solution, and invert the tube to mix three to five times.
6. Incubate on a Nutator for 30 min at room temperature.
7. Spin down, and then place the sample on a magnetic stand until solution is clear.
8. Resuspend in 500 μL wash buffer #1 on a vortex mixer.
9. Incubate for 15 min at room temperature.
10. Spin down, and then place the sample on a magnetic stand until solution is clear.

11. Wash beads three times:
    (a) Resuspend in 500 μL of 65 °C prewarmed SureSelect Wash Buffer #2 and vortex.
    (b) Incubate for 10 min at 65 °C in a recirculating water bath or heat block.
    (c) Spin down, and then place the sample on a magnetic stand until solution is clear.
    (d) Remove supernatant.
12. Mix beads in 50 μL of Elution buffer (see Note 7).
13. Incubate for 10 min at room temperature.
14. Spin down, and then place the sample on a magnetic stand until solution is clear.
15. Transfer the supernatant into a new 1.5 mL tube; add 50 μL of neutralization buffer, and mix.

3.2.3 Purify Sample

1. Add 180 μL of XP beads to a 1.5 mL LoBind tube and add 100 μL of captured DNA. Mix well and incubate for 5 min.
2. Spin down, and then place the sample on a magnetic stand until solution is clear.
3. Wash with 0.5 mL of 70 % ethanol three times.
4. Dry the samples at 37 °C for 5 min.
5. Add 30 μL of nuclease-free water and incubate for 2 min at room temperature.
6. Spin down, and then place the sample on a magnetic stand until solution is clear.
7. Remove supernatant (30 μL) to a fresh 1.5 mL LoBind tube.

3.2.4 PCR Amplification

1. Prepare the following master mix on ice. Mix thoroughly by pipetting up and down.

2× Phusion Master Mix with HF Buffer	100 μL
PCR grade water	62 μL
Primer 1, 100 μM	4 μL
Primer 2, 100 μM	4 μL
Total	170 μL

2. Aliquot 15 μL of Captured DNA library template into two PCR tubes.
3. Add 85 μL of the PCR Amplification Master Mix from step 1 into the two tubes from step 2. Mix thoroughly.
4. Amplify in a thermal cycler using the following protocol:

   98 °C for 30 s.

Ten cycles of:

(a) 98 °C for 10 s.

(b) 60 °C for 30 s.

(c) 72 °C for 30 s.

72 °C for 10 s.

Hold at 4 °C.

Pool the two reactions from each amplified captured DNA sample into one 1.5 mL tube.

5. Add 100 μL of well-mixed AMPure XP Beads to the tube. Mix up and down using pipette, changing tip after each sample.
6. Incubate the tube at room temperature for 15 min.
7. Place the tube on the magnetic stand at room temperature for 2 min.
8. Remove and discard 195 μL of supernatant. Some liquid may remain in each well. Change the tip after each sample.
9. With the tube remaining on the magnetic stand, add 500 μL of freshly prepared 80 % EtOH without disturbing the beads.
10. Incubate the tube at room temperature for at least 30 s, then remove and discard all supernatant.
11. Repeat previous two steps once for a total of two 80 % EtOH washes.
12. Let the tube dry at room temperature for 15 min, then remove from the magnetic stand.
13. Resuspend with 32.5 μL Resuspension Buffer.
14. Mix up and down using pipette.
15. Incubate at room temperature for 2 min.
16. Place the tube on the magnetic stand at room temperature for 2 min.
17. Transfer 30 μL of the clear supernatant to a new tube.

3.3 Sequencing

The enriched sample is now ready for Illumina sequencing.

3.4 Sequencing Analysis

We recommend the current best practices for variant detection developed by the Broad Institute for analysis of the 1,000 genomes project (http://www.broadinstitute.org/gatk/guide/workflows#1186).

1. Align raw read calls to the human reference genome (currently version hg19) using the program BWA (5).

```
bwa aln hg19.fa read1.fastq>aln_sa1.sai
bwa aln hg19.fa read2.fastq>aln_sa2.sai
bwa sampe hg19.fa aln_sa1.sai aln_sa2.sai read1.fastq read2.fastq>aln.sam
```

2. Convert SAM format to sorted BAM format.

   samtools view -bS -o aln.bam aln.sam

   samtools sort aln.bam aln.sorted

   samtools index aln.sorted.bam

3. Remove PCR duplicates using samtools and index the alignment.

   samtools rmdup aln.sorted.bam dedup.bam

   samtools index dedup.bam

4. Realign each subject's reads based on known indels in the human genome and recalibrate quality scores using the Genome Analysis Toolkit (GATK) (6, 7).

   java -jar GenomeAnalysisTK/GenomeAnalysisTK.jar

   – T RealignerTargetCreator

   – R hg19.fa

   – I dedup.bam

   – o outputs.intervals

   java -jar GenomeAnalysisTK/GenomeAnalysisTK.jar

   – T IndelRealigner

   – R hg19.fa

   – targetIntervals outputs.intervals

   – I dedup.bam

   – o realigned.bam

5. Add Read Group information to the BAM file for downstream SNP calling.

   java -jar picard-tools/AddOrReplaceReadGroups.jar

   – I = realigned.bam

   – O = realignedNewRG.bam

   – SORT_ORDER = coordinate

   – RGID = Flowcell_ID

   – RGLB = Library_ID

   – RGPL = Illumina

   – RGSM = Sample_ID

   – RGPU = Barcode_ID

   – CREATE_INDEX = True

6. Determine coverage and capture performance.

   java -jar picard-tools/CalculateHsMetrics.jar

   – BAIT_INTERVALS = bait.bed

   – TARGET_INTERVALS = target.bed

- INPUT = realigned.bam
- OUTPUT = picardHsMetrics.txt
- VALIDATION_STRINGENCY = LENIENT
- REFERENCE_SEQUENCE = hg19.fa

7. Call SNPs for subjects with 30× coverage at more than 90 % of the target region (i.e., the olfactory sub-genome). Variants are called for the entire set of subjects simultaneously to increase accuracy (see Note 8).

   java -Xmx8g -jar/home/Jason/GenomeAnalysisTK/GenomeAnalysisTK.jar

   - T UnifiedGenotyper
   - R hg19.fa
   - glm BOTH
   - I listOfSamples.list
   - o calls.vcf

8. This will produce a variant call format (vcf) file for downstream analysis. A common follow-up analysis is to measure the association between genetic variants and psychophysical ratings using a Wald test implemented in PLINK (8) (see Note 9).

4 Notes

1. See Roland & Reich (2) for an alternative bead mixture that may save reagent costs.
2. To avoid cross-contamination in multi-well plates, avoid vortexing the plates. Mix using a multichannel pipette. In addition, plates should be briefly spun in a swing-bucket centrifuge if solution adheres to the walls of the well.
3. It is important to quantify the concentration of the input DNA with a direct double-stranded DNA-specific binding method, such as PicoGreen (Invitrogen) or Qubit (Invitrogen). Spectrophotometric methods (e.g., NanoDrop) should be avoided throughout this protocol, as they do not have sufficient accuracy.
4. Removal of the eluted sample from the reaction vessel may not be strictly necessary. Fisher et al. (9) reported that beads should be left in the reaction vessel to reduce sample transfer steps.
5. Not all regions can be captured successfully, due to duplications in some OR genes leading to high sequence similarity that cannot be resolved.
6. Evaporation during the hybridization protocol will alter the salt concentration, and therefore the stringency of binding.

Incubate 27 μL of buffer at 65 °C for the intended time period as a test. Include buffer in each well of the plate you intend to use, as evaporation may vary by position. Evaporation should not exceed 3–4 μL.

7. Fisher et al. (9) report that the PCR mixture for the subsequent step can be added directly to the bead mixture.
8. You may wish to use the "–dcov 50" option to limit the depth of coverage used to the first 50× to speed up SNP calling.
9. Although we see high concordance in SNP calling between Illumina sequencing of targeted capture products and Sanger sequencing, it is important to validate a random sample of loci using an alternate sequencing method.

References

1. Glenn TC (2011) Field guide to next-generation DNA sequencers. Mol Ecol Resour 11: 759–769
2. Rohland N, Reich D (2012) Cost-effective, high-throughput DNA sequencing libraries for multiplexed target capture. Genome Res 22:939–946
3. Mamanova L, Coffey AJ, Scott CE et al (2010) Target-enrichment strategies for next-generation sequencing. Nat Methods 7:111–118
4. Gnirke A, Melnikov A, Maguire J et al (2009) Solution hybrid selection with ultra-long oligonucleotides for massively parallel targeted sequencing. Nat Biotechnol 27:182–189
5. Li H, Durbin R (2009) Fast and accurate short read alignment with burrows-wheeler transform. Bioinformatics 25:1754 1760
6. DePristo MA, Banks E, Poplin R et al (2011) A framework for variation discovery and genotyping using next-generation DNA sequencing data. Nat Genet 43:491–498
7. McKenna A, Hanna M, Banks E et al (2010) The genome analysis toolkit: a mapreduce framework for analyzing next-generation DNA sequencing data. Genome Res 20: 1297–1303
8. Purcell S, Neale B, Todd-Brown K et al (2007) Plink: a tool set for whole-genome association and population-based linkage analyses. Am J Hum Genet 81:559–575
9. Fisher S, Barry A, Abreu J et al (2011) A scalable, fully automated process for construction of sequence-ready human exome targeted capture libraries. Genome Biol 12:R1

Chapter 11

Deciphering Activation of Olfactory Receptors Using Heterologous Expression in *Saccharomyces cerevisiae* and Bioluminescence Resonance Energy Transfer

Guenhaël Sanz and Edith Pajot-Augy

Abstract

Hetero- and homo-oligomerization of G protein-coupled receptors (GPCRs) has been addressed in the past years using various approaches such as co-immunoprecipitation, fluorescence resonance energy transfer and bioluminescence resonance energy transfer (BRET). Here, we report the methodological details from a previously published study to investigate the relationships between oligomerization and activation states of olfactory receptors (ORs). This methodology combines heterologous expression of ORs in *Saccharomyces cerevisiae* and BRET assays on membrane fractions, in particular, upon odorant stimulation. We have demonstrated that ORs constitutively homodimerize at the plasma membrane and that high odorant concentrations promote a conformational change of the dimer, which becomes inactive. We proposed a model in which one odorant molecule binding the dimer would induce activation, while two odorant molecules, each binding one protomer of the dimer, would blunt signaling.

Key words Olfactory receptor, BRET, Heterologous expression, Homodimerization, GPCR

1 Introduction

The dimerization of olfactory receptors (ORs) with other G protein-coupled receptors (GPCRs) (1, 2) or membrane proteins (3) has already been reported. Our study, however, was the first to establish OR homodimerization and its role in ORs activation at various ligand concentrations (4). Most notably, we demonstrated that high doses of odorants induce an inactive conformation of the OR dimers. To this end, we used optimized heterologous expression of ORs in *Saccharomyces cerevisiae* and bioluminescence resonance energy transfer (BRET), a biophysical technology commonly employed to monitor GPCR oligomerization and ligand-mediated conformational changes in oligomers (5–7).

BRET studies are conducted more commonly during heterologous expression of GPCRs in mammalian cells than in

Chiquito J. Crasto (ed.), *Olfactory Receptors: Methods and Protocols*, Methods in Molecular Biology, vol. 1003,
DOI 10.1007/978-1-62703-377-0_11, © Springer Science+Business Media, LLC 2013

yeasts (8). Among the various BRET techniques, varying donors, acceptors, and substrates, we chose the BRET1. BRET1 uses Rluc as the donor, enhanced YFP (EYFP) as the acceptor, and coelenterazine h as the substrate. In this case, however, the overlap between the donor and acceptor emission spectra is responsible for a background signal. Taken into account this background noise, the high level of bioluminescent emission obtained using coelenterazine h allows the detection of a BRET signal while expressing proteins of interest at physiological levels. This minimizes nonspecific interactions due to over-expression. BRET2, whose donor and acceptor emission peaks are well separated, results in a low induced background. But the low amplitude of BRET2 signals makes it necessary to over-express proteins, which is detrimental to interaction specificity.

Though an adequate expression level of ORs is reached in our yeast expression system, which is not an easy task for receptors of this type, the system is difficult to control. Even using BRET1, it is necessary to control the specificity of the measured interactions. For that purpose, BRET measurements were performed as a function of the amount of acceptor relative to the donor. Specific BRET depends on the acceptor/donor ratio and saturates with increasing ratios, a tendency we have observed. Moreover, a further proof of specificity was provided by measuring the variation of the BRET level upon ligand stimulation.

2 Materials

2.1 Yeast Growth and Transformation

1. *S. cerevisiae* strain MC18 (MATα gpa1::lacZ (LEU2) ade2-1 his3-11, 15 leu2-3112 trp1-1 ura3-1 can1–100) transformed with pRGP-Golf (see Note 1).
2. Growth medium A: yeast nitrogen base, synthetic drop-out CSM media without HIS, LEU, TRP, URA (Bio101), 40 mg/L adenine, 40 mg/L tryptophane, 20 mg/L uracil, supplemented with 2 % glucose.
3. Growth and selection medium B for transformed yeast: medium A depleted of TRP and/or URA.
4. Induction media: medium B supplemented with 3 % lactate, then medium B supplemented with 2 % galactose.
5. Transformation solutions: 50 % PEG 4,000 in H_2O Milli-Q, 1 M lithium acetate in H_2O Milli-Q, 10 mg/mL salmon sperm DNA.
6. Medium for one transformation (346 μL final): 240 μL PEG 4,000, 36 μL lithium acetate, 65 μL H_2O Milli-Q, 5 μL of preheated (95 °C, 10 min) salmon sperm DNA.

2.2 Yeast Expression Plasmids

1. pESC-URA-cmyc-OR1740-Rluc: pESC-URA (Stratagene) modified to express upon galactose induction the human OR1740 receptor tagged by the c-myc epitope at its N-terminal end and fused to Rluc at its C-terminal end.
2. pESC-TRP-cmyc-OR1740-EYFP: pESC-TRP (Stratagene) modified to express upon galactose induction the human OR1740 receptor tagged by the c-myc epitope at its N-terminal end and fused to EYFP at its C-terminal end (see Note 2).

2.3 Crude Membrane Preparation

1. Lysis buffer A: 50 mM Tris–HCl pH 7.5, 1 mM EDTA, 250 mM sorbitol, 0.1 mM PMSF, complete protease inhibitor mixture (Roche Applied Science).
2. Resuspension buffer B: 10 mM Tris–HCl pH 7.5, 1 mM EDTA, 1 mM DTT, complete protease inhibitor mixture.
3. BCA reagent.

2.4 SDS-Polyacrylamide Gel Electrophoresis

1. Mini Protean II™ (Bio-Rad).
2. 1.5 M Tris–HCl pH 8.8 (for separating gel). Stored at room temperature.
3. 1 M Tris–HCl pH 6.8 (for stacking gel). Stored at room temperature.
4. 10 % SDS. Stored at room temperature.
5. 30 % acrylamide: 0.8 % bis-acrylamide (37.5:1, Protogel) (since it is a neurotoxin when unpolymerized, use under a sorbonne).
6. *N,N,N,N'*-Tetramethyl-ethylenediamine (TEMED). Store at 4 °C.
7. Ammonium persulfate: prepare extemporaneously a 10 % solution in H_2O Milli-Q. Aliquots can be frozen at −20 °C (in single use).
8. Running buffer 5×: 15.1 g/L Tris-Base, 94 g/L glycine, 0.5 % (w/v) SDS. Store at room temperature.
9. Prestained protein ladder (Euromedex). Store at −20 °C.
10. Loading buffer (Laemmli) 5×: 3.125 mL 1 M Tris–HCl pH 6.8, 0.5 mL β-mercaptoethanol, 0.125 mL 1 % (w/v) bromophenol blue, 5 mL glycerol, 10 % (w/v) SDS, qsp 10 mL with H_2O Milli-Q.

2.5 Immunoblotting

1. Mini Trans-Blot Electrophoretic Transfer Cell (Bio-Rad).
2. Transfer buffer 10×: 30.3 g/L Tris-Base, 144.13 g/L glycine, qsp 1 L with H_2O Milli-Q. Store at room temperature. Extemporaneously mix 100 mL of the 10× buffer with 100 mL of absolute ethanol and complete to 1 L with H_2O Milli-Q.

3. Immobilon-P transfer membranes (Millipore): before use, soak the membranes 15 s in methanol and 2 min in H_2O Milli-Q.
4. 3MM paper.
5. Ponceau red 0.2 % (w/v).
6. Primary antibody: anti-c-myc (Roche): 1 mg/mL in PBS and storage at −20 °C in aliquots to avoid repeated freezing and thawing (see Note 3).
7. Secondary antibody: peroxidase-conjugated anti-mouse IgG (Sigma).
8. Blocking buffer: PBS, 0.1 % Tween 20, 4.5 % nonfat milk.
9. Hybridization buffer: PBS, 0.1 % Tween 20, 4.5 % nonfat milk.
10. Washing buffers: PBS, 0.1 % Tween 20, 0.5 % or 0.1 % nonfat milk.
11. Enhanced chemiluminescent (ECL) Plus reagent (PerkinElmer Life Sciences).
12. BioMax Light Films (Kodak).

2.6 BRET Assays

1. BRET buffer: PBS, 0.01 % (w/v) $MgSO_4$, 0.1 % (w/v) glucose.
2. Coelenterazine h (Promega).
3. TriStar LB 941 multimode reader (Berthold) equipped with emission filters at 485 and 530 nm.
4. White 96-well microplates.

2.7 Odorants

1. 2 mL glass vials.
2. Odorants at the highest available purity.
3. DMSO (see Note 4).

3 Methods

3.1 Yeast Growth

1. Yeasts were first plated on 2 % agar in medium A or B and incubated 48 h at 30 °C.
2. 3 mL of medium A or B was inoculated with a yeast colony and incubated 24 h at 30 °C and 150 rpm.
3. Yeasts were then grown in medium A or B with an absorbance not exceeding 2. In general, inoculate medium A or B with 10 % yeasts at an absorbance between 1 and 2.

3.2 Plasmids Transformation in Yeast

1. Each transformation requires 5 mL of yeasts with an absorbance between 1 and 2. Prepare the volume of yeasts according to the number of transformations.
2. Spin yeasts 5 min at $1500\times g$ and 4 °C to remove growth medium.

3. Resuspend the pellet in the same volume of H_2O Milli-Q to wash yeasts. Spin again to remove water (see Note 5).
4. Resuspend the pellet in 1 mL H_2O Milli-Q and transfer to a 1.5 mL tube. Spin again (15 s at 9,300×*g*) to remove water (see Note 5).
5. Resuspend the pellet in 500 μL 0.1 M lithium acetate and incubate 15 min at 30 °C (see Note 5).
6. During this time, heat salmon sperm DNA at 95 °C for 10 min. Cool and maintain it on ice.
7. Prepare transformation medium according to the number of transformations as follows: mix by vortexing PEG 4,000, lithium acetate, and H_2O Milli-Q and then add pretreated salmon sperm DNA.
8. Spin yeasts 15 s at 9,300×*g* to remove lithium acetate.
9. Resuspend the pellet in the transformation medium (see Note 5) and aliquot according to the number of transformations in 1 mL tubes. In each tube, add 4 μg of plasmidic DNA to transform in 4 μL of H_2O Milli-Q or 4 μL H_2O Milli-Q for controls.
10. Incubate yeasts 30 min at 30 °C, then 20 min at 42 °C.
11. Spin (15 s at 9,300×*g*) to remove transformation medium.
12. Resuspend each pellet in 100 μL H_2O Milli-Q and plate 90 μL and 10 μL of yeasts on 2 % agar in medium B. Incubate 48 h at 30 °C.

3.3 Induction of OR Expression in Yeast

1. Start with colony of yeasts transformed with the expression vector pESC-URA-cmyc-OR1740-Rluc, pESC-TRP-cmyc-OR1740-EYFP, or both.
2. Inoculate 3 mL of medium B with the colony. Yeasts are grown in at least 20 mL of medium B until they reach an absorbance between 1 and 2.
3. Spin yeasts 5 min at 1,500×*g* and 4 °C to remove growth medium.
4. Resuspend the pellet in 20 mL of H_2O Milli-Q to wash yeasts. Spin again to remove water.
5. Resuspend the pellet in 20 mL of medium B with 3 % lactate. Incubate 6 h at 30 °C and 150 rpm.
6. Spin to remove the medium and resuspend the pellet in 20 mL of medium B with 2 % galactose. Adjust absorbance to 1 with medium (do not exceed a culture volume of 50 mL) and incubate yeasts in an Erlenmeyer at 15 °C and 150 rpm during 108 h (see Note 6). 48 h after the beginning of induction, readjust absorbance to 1 by adding medium.

3.4 Crude Membrane Preparation for BRET Experiments

1. Start with at least 40 mL of yeast culture (absorbance around 5).
2. Spin yeasts 5 min at 1,500 × *g* and 4 °C to remove growth medium.
3. Wash twice the pellet of yeasts with 4 mL of ice-cold H_2O Milli-Q and then resuspend it in 4 mL of lysis buffer prepared extemporaneously and kept on ice.
4. Aliquot in 2 mL tubes (do not exceed 1 mL).
5. Add glass beads of 0.5 mm diameter (50 % of the initial volume). Vortex each tube six times at full speed during 1 min and keep them 1 min on ice between each vortexing.
6. Collect supernatants and transfer them in 15 mL tubes.
7. Wash the glass beads twice with 400 mL of lysis buffer, with brief vortexing. Transfer the supernatants into 15 mL tubes.
8. Spin tubes 10 min at 2,400 × *g* to remove unbroken yeasts and fragments.
9. Transfer supernatants in tubes for ultra-centrifugation with a Beckman TLA 100.3 rotor during 40 min at 4 °C and 55,000 × *g*.
10. Resuspend the pellets in 500 μL of buffer B and transfer in 1.5 mL tubes. Homogenize samples with a silicon dounce adapted for 1.5 mL tubes. Store these membrane fractions at −80 °C.
11. Quantify protein content using BCA reagent (Pierce) according to the manufacturer's instructions and using bovine serum albumin as a standard.

3.5 SDS-PAGE

1. Prepare 15 mL of 10 % SDS-polyacrylamide gel (SDS-PAGE) (separating gel) by mixing H_2O Milli-Q (5.7 mL), acrylamide/bis-acrylamide solution (5.1 mL), 1.5 M Tris–HCl pH 8.8 (3.9 mL), 10 % SDS (150 μL), 10 % ammonium persulfate (150 μL), and TEMED (6 μL).
2. Cast the polymerizing gel in the Mini Protean system. Overlay with H_2O Milli-Q until the polymerized (about 30 min).
3. Prepare 8 mL of stacking gel by mixing H_2O Milli-Q (5.5 mL), acrylamide/bis-acrylamide solution (1.3 mL), 1 M Tris–HCl pH 6.8 (1 mL), 10 % SDS (80 μL), 10 % ammonium persulfate (80 μL), and TEMED (8 μL).
4. Remove the water above the separating gel and cast the stacking gel. Insert the comb. The stacking gel should polymerize within 30 min.
5. Prepare 1 L of running buffer.
6. Remove the comb from the stacking gel and wash the wells with running buffer using a thin cone.

7. Add running buffer in the Mini Protean system so that the gel lower and upper parts are immersed.
8. Prepare samples by mixing crude membranes corresponding to 5 µg of protein content with loading buffer. Do not boil samples to avoid aggregation.
9. Load samples and prestained protein ladder.
10. Run at 30 mA through the stacking gel, then at 40 mA through the separating gel (constant intensity).

3.6 Immunoblotting

1. Samples separated by SDS-PAGE are electro-transferred onto Immobilon-P transfer membranes using the Mini Trans-Blot Electrophoretic Transfer Cell system according to the manufacturer's instructions. Before use, the membranes are cut precisely to the size of the separating gel to avoid current leakage during transfer, and soaked in methanol and then H_2O Milli-Q. Mark the orientation and side of the membrane (cutting a corner is fine). Carefully cut the stacking gel from the separating gel that will be electro-transferred. Take care to soak 3MM papers, membrane and pads with transfer buffer. Place, in that order, a pad, two 3MM papers, the separating gel, a membrane, two 3MM papers and a pad, taking care not trap to any bubbles between the separating gel and membrane, and 3MM papers. Make sure to place the resulting sandwich in the correct orientation in the electro-transfer system.
2. Place the electro-transfer system in a polystyrene box with ice to avoid heating. Run the transfer of proteins to the membrane at 200 mA (constant intensity) for 3 h.
3. When the transfer is completed, disassemble the sandwich and recover the membrane.
4. At this stage, colored proteins of the prestained protein ladder should be clearly visible on the membrane. Other proteins can be revealed by soaking the membrane in Ponceau red. Wash the membrane in a bath of water to see the electro-transferred proteins. Before proceeding to the next stages of the protocol, remove the red coloration by washing the membrane in a PBS bath during a few minutes.
5. Incubate the membrane in 50 mL blocking buffer for 1 h at room temperature on a rocking platform.
6. Remove the blocking buffer and incubate the membrane in 3 mL hybridization buffer containing the diluted (1:1,000) primary antibody, overnight at 4 °C on a rotating platform. Use a small sealed plastic bag to incubate the membrane in order to reduce the volume of antibody.

Fig. 1 Expression of the OR1740 receptor fused to Rluc or EYFP. Immunoblot of crude membranes from yeasts expressing OR1740-Rluc alone (*lane 1*) or co-expressing OR1740-Rluc and OR1740-EYFP (*lane 2*). Yeast expression of the receptors was induced during 108 h at 15 °C. Receptors, tagged with a c-myc epitope at their N-terminal end, were revealed using an anti-c-myc antibody

7. Wash the membrane four times for 15 min each in 50 mL washing buffer (0.5 % milk) at room temperature on a rocking platform.
8. Incubate the membrane in 3 mL hybridization buffer containing the diluted (1:2,000) secondary antibody, for 1 h at room temperature on a rotating platform. Use a small sealed plastic bag to incubate the membrane. Longer incubation at that temperature would result in increased background.
9. Wash the membrane four times for 15 min each in 50 mL washing buffer (0.1 % milk) at room temperature on a rocking platform.
10. Mix together 1 mL of each ECL reagent and immediately cover the entire surface of the membrane. One minute later, the reagent is delicately drained from the membrane by a corner on a tissue, and the membrane is placed in a sealed light plastic bag. In a dark room, a BioMax Light film is exposed to the membrane in an X-ray film cassette for a suitable exposure time (typically 1 min or less). The film is revealed. An example of this result is shown in Fig. 1.

3.7 OR Homodimerization Assessment by BRET

1. BRET measurements are performed using yeast crude membranes carrying the OR fused to EYFP (OR-EYFP) and/or the OR fused to Rluc (OR-Rluc). For each measurement with membranes carrying both OR-Rluc and OR-EYFP, include a measurement with membranes carrying only OR-Rluc. Use crude membranes for which the expression of OR-EYFP and/or OR-Rluc by immunoblotting is checked (see Note 7).
2. For each test, use a quantity of crude membranes corresponding to 10 μg of protein content (typically 1–5 μL). Place samples in a 96-well microplate and complete to 25 μL with BRET buffer. Just before measurement, 25 μL of BRET buffer containing 10 μM coelenterazine h are added and mixed rapidly by pipetting to obtain a final concentration of 5 μM. Once the coelenterazine h is added, immediately measure emissions at

530 and 485 nm (1 s each), and repeat measurements each minute for at least 5 min to monitor the evolution of the BRET signal. The BRET signal is then expressed as the ratio of emission at 530 nm to 485 nm, and results are normalized by subtracting the ratio measured with OR-Rluc alone to the one measured with both OR-Rluc and OR-EYFP: mBRET = (530/485 (OR-Rluc + OR-EYFP)–530/485 (OR-Rluc)) × 1,000. These measurements can be performed either using automatic injections of coelenterazine h or by adding it manually just before measurements, not exceeding four wells at a time to ensure reproducibility.

- To ensure specificity of the BRET signal observed, a BRET saturation assay can be performed, which consists in measuring the BRET signal for various acceptor (OR-EYFP)/donor (OR-Rluc) ratios, with a relatively constant donor amount. For that purpose, crude membranes from various yeast clones co-expressing OR-Rluc and OR-EYFP should be used, because they naturally exhibit varying relative amounts of the two expressed receptors (to be evaluated by immunoblotting using anti-c-myc antibody) (see Fig. 1 for an example). Check that luminescence emission at 530 nm is similar for the various samples tested. Then, measure the BRET signal and plot it against the corresponding OR-EYFP/OR-Rluc ratio value for the various clones. The curve obtained should display a tendency towards saturation if the BRET signal corresponds to specific interactions of ORs and not to random, nonspecific clustering of ORs at high densities (which can be the case when ORs are over-expressed).

3. Previously published results obtained with the OR1740 receptor are shown in Fig. 2 (4). They demonstrate that BRET occurs specifically, thus that the OR1740 receptor indeed forms homodimers.

3.8 BRET Modulation upon Odorant Ligand Stimulation

1. Odorants solutions should be prepared in glass vials with a cap, under a sorbonne and with gloves (change gloves for each odorant). Dilute pure odorant first in DMSO to obtain a stock solution of at least 10^{-1} M, then 1,000-fold in BRET buffer, and in BRET buffer for further dilutions.
2. Perform BRET experiments in the presence of various amounts of odorant ligand. Add odorants in the BRET buffer used to dilute coelenterazine h in order to obtain the desired final concentration in the well (see Note 8). Try to work rapidly when manipulating odorants because of their volatility. It is also essential to perform BRET measurements immediately after odorant addition, since a rapid and transient variation of

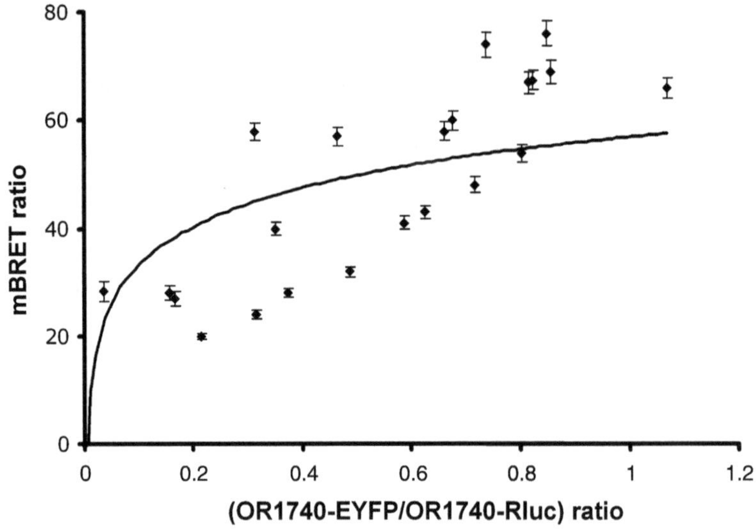

Fig. 2 OR1740 homodimerization assessment by BRET. BRET level was measured from crude membranes of yeast cells co-expressing OR1740-Rluc (donor) and OR1740-EYFP (acceptor) or expressing OR1740-Rluc alone. Clones with variable acceptor (OR1740-EYFP) to donor (OR1740-Rluc) ratios were used. Data are expressed as the normalized BRET ratio and plotted as a function of the acceptor to donor ratio, estimated from immunoblots (an example is shown in Fig. 1). Data are representative of at least three independent experiments. Figure redrawn from (4)

the BRET signal is expected, due to activation of the receptor. To control the variation of the BRET signal that is specifically due to binding of the odorant agonist to the receptor, odorants unrelated to the receptor, but with a structure relatively similar to that of the agonist tested, should also be used as negative controls. Also ensure that the highest quantity of DMSO used to dilute odorants has no effect on the BRET signal obtained with BRET buffer only. Our results demonstrated that the preexisting BRET signal increases upon odorant ligand stimulation, which supports a conformational change of the OR dimer upon ligand binding. Furthermore, modulation of the BRET signal upon stimulation with increasing amounts of odorant displays a bell-shaped curve. Comparison of these results with the bell-shaped curve of the OR functional response with increasing doses of odorant (9) suggests that high odorant ligand concentrations mediate a conformational rearrangement of the OR dimer that is unfavorable to activation (see Fig. 3).

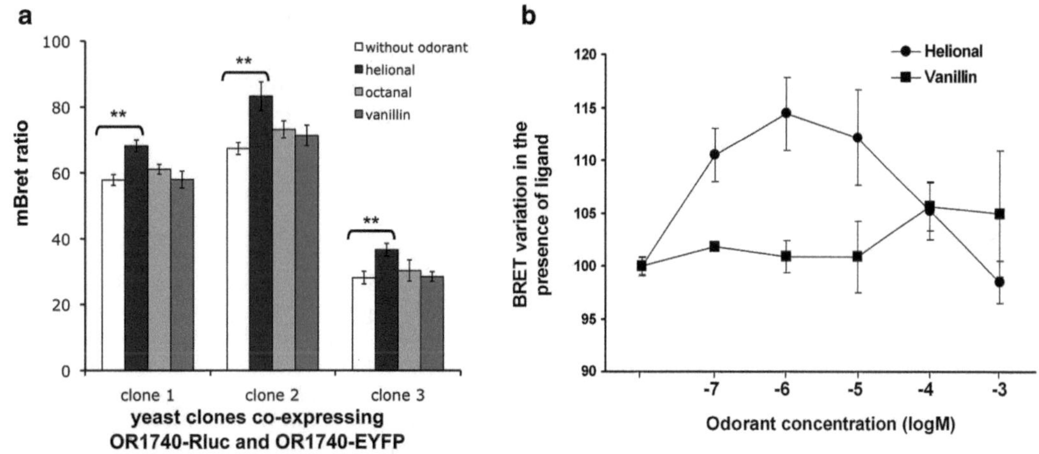

Fig. 3 BRET modulation upon odorant stimulation. (**a**) Crude membranes from three different clones co-expressing OR1740-Rluc and OR1740-EYFP were used to perform BRET assays without or with odorants (helional as an OR1740 agonist, octanal and vanillin as negative controls). (**) shows significant differences estimated by Student test ($P < 0.05$). Data are representative of three independent experiments. (**b**) BRET level variation upon OR1740 stimulation with various helional concentrations. BRET measurements were performed using crude membranes from yeast cells co-expressing OR1740-Rluc and OR1740-EYFP or expressing OR1740-Rluc alone. Receptors were stimulated with various concentrations of helional. Results are expressed as the relative variation of BRET level at the various odorant concentrations. Data are representative of three independent experiments. Figure redrawn from (4)

4 Notes

1. pRGP-Golf provides the yeast MC18 strain with the ability to grow without histidine complementation (10).
2. Modified vectors can be provided upon request. Constructions were already detailed in Wade 2011.
3. The anti-c-myc antibody allows the detection of ORs tagged with the c-myc epitope at their N-terminal end.
4. Use DMSO sold in 5 mL vials. Once opened, keep tightly closed, away from light and for less than 2 weeks.
5. At this stage of the protocol, yeasts to be used for various transformations can be pooled.
6. Induction of OR expression was maintained for 108 h because we demonstrated that this duration is necessary for targeting ORs to the plasma membrane (4).
7. BRET measurements can be performed on disrupted yeast cells instead of using crude membranes, but signal amplitudes are more than two times lower (4).

8. Because we prefer not to use automatic injection of odorants (risk of odorant contamination of the injection system), they are added manually with coelenterazine h. So, do not exceed four wells to be measured at a time to ensure reproducibility.

References

1. Hague C, Uberti MA, Chen Z, Bush CF, Jones SV, Ressler KJ, Hall RA, Minneman KP (2004) Olfactory receptor surface expression is driven by association with the beta2-adrenergic receptor. Proc Natl Acad Sci USA 101: 13672–13676
2. Bush CF, Jones SV, Lyle AN, Minneman KP, Ressler KJ, Hall RA (2007) Specificity of olfactory receptor interactions with other G protein-coupled receptors. J Biol Chem 282: 19042–19051
3. Saito H, Kubota M, Roberts RW, Chi Q, Matsunami H (2004) RTP family members induce functional expression of mammalian odorant receptors. Cell 119:679–691
4. Wade F, Espagne A, Persuy MA, Vidic J, Monnerie R, Merola F, Pajot-Augy E, Sanz G (2011) Relationship between homo-oligomerization of a mammalian olfactory receptor and its activation state demonstrated by bioluminescence resonance energy transfer. J Biol Chem 286:15252–15259
5. Milligan G, Bouvier M (2005) Methods to monitor the quaternary structure of G protein-coupled receptors. FEBS J 272:2914–2925
6. Gandia J, Lluis C, Ferre S, Franco R, Ciruela F (2008) Light resonance energy transfer-based methods in the study of G protein-coupled receptor oligomerization. Bioessays 30: 82–89
7. Bacart J, Corbel C, Jockers R, Bach S, Couturier C (2008) The BRET technology and its application to screening assays. Biotechnol J 3:311–324
8. Gehret AU, Bajaj A, Naider F, Dumont ME (2006) Oligomerization of the yeast alpha-factor receptor: implication for dominant negative effects of mutant receptors. J Biol Chem 281:20698–20714
9. Vidic J, Grosclaude J, Monnerie R, Persuy MA, Badonnel K, Baly C, Caillol M, Briand L, Salesse R, Pajot-Augy E (2008) On a chip demonstration of a functional role for odorant binding protein in the preservation of olfactory receptor activity at high odorant concentration. Lab Chip 8:678–688
10. Crowe ML, Perry BN, Connerton IF (2000) Golf complements a GPA1 null mutation in *Saccharomyces cerevisiae* and functionally couples to the STE2 pheromone receptor. J Recept Signal Transduct Res 20:61–73

Chapter 12

Olfactory Receptors in Human Airway Epithelia

Xiaoling Gu and Yehuda Ben-Shahar

Abstract

Olfactory receptors (OR) represent one of the largest gene families in the human genome. In spite of a significant progress in deciphering the physiological functions of olfactory receptors, how the majority of these G-protein-coupled receptors are activated is still mostly a mystery. Consequently, for the majority of OR genes there are currently no assigned physiological or behavioral functions. Deciphering ligand specificities and physiological significance of human ORs is important for understanding how the human olfactory genome encodes odors, and how such odors drive human behavior in health and disease. Although OR genes were originally thought to be restricted to the olfactory epithelium, several recent studies indicated that some members of the OR family might be acting outside the canonical chemosensory system. In a recent study, we have shown that the human airway epithelial cells can also act as chemosensory cells by directly sensing the inhalation of noxious bitter compounds, which can lead to increased mucociliary clearance, and hence may serve as a protective mechanism against inhaled toxins and microorganisms. Whether the airway epithelium can detect chemicals via other sensory pathways has not been reported to date. As a step in this direction, we describe methods for studying the cellular and subcellular localization of olfactory receptor proteins and mRNAs in human airways in both primary in vitro cultures and tissue sections.

Key words Olfactory receptors, Pulmonary epithelium, Confocal immunofluorescence, Homo sapiens, Primate, Lung

1 Introduction

The rat olfactory receptor (OR) family was first described in 1991 (1). Since then, many additional publications established members of the family as the principle molecular olfactory receptors in the vertebrate main olfactory epithelium where they detect inhaled volatile chemicals (e.g., (2–5)). The human genome contains at least 339 genes that encode for functional OR proteins and 297 OR pseudogenes (3). Although OR genes are clearly playing an olfactory role, the vast majority of the members of this superfamily are orphan receptors, i.e., the specific ligand(s) that activate them are unknown. Nevertheless, decoding the ligand specificity of

Chiquito J. Crasto (ed.), *Olfactory Receptors: Methods and Protocols*, Methods in Molecular Biology, vol. 1003, DOI 10.1007/978-1-62703-377-0_12, © Springer Science+Business Media, LLC 2013

human ORs is important for understanding how the human olfactory genome encodes odors, and how such odors drive human behavior in health and disease. Furthermore, natural genetic variations in the ability of individuals to respond physiologically and behaviorally to specific odors can teach us much about the evolution of our own species (6).

Although the main olfactory epithelium is thought to be the sole organ for the detection of volatile chemicals, some OR genes seem to be acting outside the canonical chemosensory system. Examples include the role of OR1D2 in the chemotactic response of human sperm to Bourgeonal (7), and the presence of OR genes in the gut epithelium (8).

We have recently demonstrated that the human airway epithelium can act as a chemosensory organ. Ciliated airway cells express members of bitter taste receptor family and are responsive to bitter compounds (9). During our investigation of human airways taste functions we also discovered that several OR genes are present in these tissues. Since primary airway cultures are readily available, this finding suggests that the human airway epithelium could serve as an excellent in vitro model for studying human olfactory functions. Here we focus on methods for describing the cellular and subcellular localization of olfactory receptor proteins and mRNAs in human airways in both primary in vitro cultures and tissue sections (Fig. 1). These methods can be easily modified for studies of olfactory receptors in other mammalian species.

2 Materials

2.1 Immunofluorescence

1. Xylene.
2. 100 % Ethanol.
3. 95 % Ethanol.
4. Phosphate-buffered saline (PBS): Dissolve the following components in 800 ml distilled H_2O. (8 g of NaCl, 0.2 g of KCl, 1.44 g of Na_2HPO_4, 0.24 g of KH_2PO_4), and adjust pH to 7.4 with HCl, add distilled H_2O to final volume 1 l. Autoclave and store the solution at room temperature.
5. Paraformaldehyde (PFA, Sigma): 4 % PFA was freshly prepared with PBS, aliquoted and stored at −20 °C.
6. SuperBlock blocking buffer (Thermo Scientific).
7. VECTASHIELD Hard set mounting medium with DAPI (Vector Laboratories).
8. Antigen Unmasking solution (Vector Laboratories).
9. PBS with Triton (PBST): Add 0.3 ml Triton X-100 to 1 l PBS.

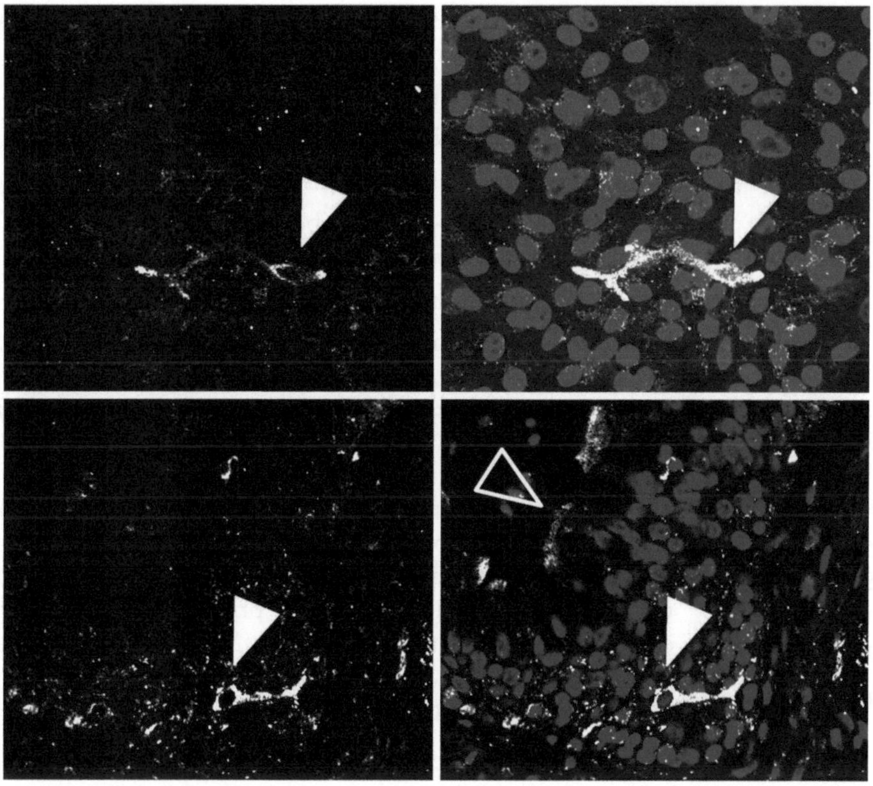

Fig. 1 Human olfactory receptor 2H3 (OR2H3) expression in airway tissues. (**a**) Pulmonary olfactory cell in human primary airway culture. (**b**) Pulmonary olfactory cell in human airway paraffin section. *Left panel*, anti-OR2H3 (Abcam); *right panel*, anti-OR2H3 channel merged with DAPI (nuclear stain). *White arrowheads*, cell bodies; *empty arrowhead*, ciliated lumen of a major airway. Note that the olfactory cell body is at the basal layer of the stratified epithelium

2.2 Double Immunofluorescence Staining

1. Conjugated Fab: Alexa Fluor 488-conjugated AffiniPure Fab Fragment Donkey anti-rabbit IgG (H+L) (Jackson ImmunoResearch).
2. Unconjugated Fab: AffiniPure Fab Fragment Donkey anti-rabbit IgG (H+L) (Jackson ImmunoResearch).

2.3 In Situ Hybridization

1. In vitro RNA transcription kit (e.g., MAXIscript, Ambion).
2. Hybridization solution: 50 % Formamide, 10 % Dextran Sulfate, 1× Denhardt's solution (Invitrogen), 10 mM Tris–HCl pH 7.5, 60 mM NaCl, 1 mM EDTA, 0.25 % SDS, 1 mg/ml yeast tRNA (Sigma).
3. Proteinase K (Roche): 1 mg/ml Proteinase K stock solution was prepared in 0.1 M Tris–HCl pH 7.5 and 50 mM EDTA.
4. 10× SSC (Thermo Scientific).

5. 5× MABT stock: 500 ml maleic acid pH 7.5, 750 mM NaCl, 0.5 % v/v Tween 20.
6. Formamide (Sigma).
7. DIG-labeled RNA probe.
8. Fluorescein-labeled RNA probe.
9. Anti-fluorescein-POD antibody (Roche), anti-DIG-POD antibody (Roche).
10. TSA™ plus Fluorescein System (PerkinElmer).
11. TSA™ plus Cyanine 3 System (PerkinElmer).

3 Methods

3.1 Preparation of Primary Cultures and Tissue Sections for Immunofluorescence Confocal Microscopy

3.1.1 Primary Cultures

This section assumes that primary human airway cultures are available. A detailed protocol for how to obtain such cultures from human donors is available (10). The protocol below requires the use of fully differentiated human primary airway cultures (high density of ciliated epithelial cells) in Millicell Cell Culture Inserts (Millipore). We find that this protocol works well regardless of the Millicell membrane type.

1. Wash inserts briefly with PBS (both sides).
2. Fix with 4 % paraformaldehyde in PBS for 15 min at room temperature.
3. Wash inserts twice with PBS.
4. Cover inserts with ice-cold methanol for 10 min on ice.
5. Wash twice with PBS.

3.1.2 Paraffin Sections

1. Deparaffinize slides in xylene, 10 min.
2. Wash slides in 100 % ethanol, 2× 10 min.
3. Wash slides in 90 % ethanol, 10 min.
4. Wash slides in 70 % ethanol, 10 min.
5. Rinse slides in distilled water.
6. If antigen unmasking is required, place slides in antigen unmasking solution at room temperature. Then bring slides to boil in antigen unmasking solution using a microwave for 10–20 min. Cool slides back to room temperature.
7. Wash slides twice with distilled water.
8. Wash slides with PBS.

3.2 Antibody Staining

This protocol can be used for a single or multiple protein targets, as long as the primary antibodies have been produced in different hosts. For a double staining of receptors with antibodies that were produced in a single host, please refer to Subheading 3.3.

1. Incubate slides with superblock blocking buffer for 30 min at room temperature.
2. Incubate slides with primary antibody at the appropriate dilution in superblock blocking buffer for 1–2 h at room temperature or overnight at 4 °C. For very rare targets, longer incubation time might increase specific signal.
3. Rinse slides with PBS/0.3 % Triton X-100.
4. Incubate slides with fluorochrome-conjugated secondary antibody at superblock blocking buffer for 30 min at room temperature.
5. Rinse in PBS/0.3 % Triton X-100.
6. Cover tissue with Vectashield Mounting medium and place a coverslip of appropriate size and thickness.
7. Seal coverslip by covering its edges with nail polish.
8. Examine slides immediately or store at 4 °C in dark.

3.3 Double Staining with Primary Antibodies from a Single Host Species

1. Incubate sections with superblock blocking buffer for 30 min at room temperature.
2. Incubate sections with primary antibody at appropriate dilution in superblock blocking buffer for 1–2 h at room temperature or overnight at 4 °C.
3. Wash sections with PBS/0.3 % Triton X-100.
4. Incubate with excess Alexa Fluor 488-conjugated AffiniPure Fab (2). If this results in high background, try a lower concentration of the probe-conjugated Fab followed by further blocking with excess *unconjugated* AffiniPure Fab. Since the first antigen will generate a lower signal, typically the more abundant protein should be labeled first (see Notes 6 and 7).
5. Rinse in PBS/0.3 % Triton X-100.
6. Incubate with the second primary antibody at appropriate dilution in superblock blocking buffer, followed by Probe-conjugated secondary antibody.
7. Rinse in PBS/0.3 % Triton X-100.
8. Rinse coverslip slides with Vectashield Mounting medium.
9. Seal slides by painting around edges of coverslips with nail polish. Examine slides immediately or store at 4 °C in dark.

3.4 *In Situ* Hybridization on Paraffin Sections Using TSA-Based Double Immunofluorescent Probes

This protocol is intended for the detection of two genes simultaneously by using confocal microscopy. It can be easily modified for using chromogenic detection reagents.

3.4.1 Generating Riboprobes for *In Situ* Hybridization

We use a commercial kit for generating in vitro RNA probes (riboprobes). Kits are available from multiple commercial sources. This protocol will focus on generating the template for the probe. We generate a DNA template by amplifying the probe from a cDNA template or genomic DNA. We also typically use the T7 RNA polymerase. Using T3 or SP6 polymerases can yield similar results.

1. We find that riboprobes with a length of 300–600 bp work well (1). To design primers we use the freely available PCR oligomer design tool available on the NCBI webpage (http://www.ncbi.nlm.nih.gov/tools/primer-blast/) (See Note 1).
2. RNA polymerases require a specific promoter sequences. We typically add the sequence for the T7 polymerase (TAATACGACTCACTATAGGG) to the 5′ end of each designed primer.
3. Two version of each primer should be synthesized with and without the T7 promoter sequence at the 5′ end.
4. For each template two PCR reaction should be set:
   (a) For the antisense probe use a reverse primer with a T7 promoter sequence and forward primer without the T7 promoter.
   (b) For a control sense probe use the reverse primer without the T7 promoter and the forward primer with the T7 promoter sequence.
5. Run the PCR reaction on a 1 % agarose gel and purify the expected PCR band. Sequence the PCR fragment to ensure it has amplified the correct template. This is especially important of genomic DNA was used as a template for the PCR.
6. To generate the riboprobe, mix the following component on ice:
   (a) PCR DNA template (100–200 ng).
   (b) 2 μl of 10× RNA labeling mix (DIG or fluorecein; Roche) (see Note 5).
   (c) 2 μl of 10× transcription buffer (each polymerase requires its own buffer) (see Note 2).
   (d) 2 μl of RNA polymerase (SP6, T7, or T3).
7. DEPC-treated water to bring the final volume to 20 μl. Mix briefly and incubate at 37 °C for 2 h. Add 2 μl of 0.2 M EDTA (pH 8) to stop the reaction (see Notes 3 and 4).

3.4.2 In Situ Hybridization

1. Deparaffinize slides in xylene, 10 min.
2. Wash slides in 100 % ethanol, 2× 10 min.
3. Wash slides in 90 % ethanol, 10 min.
4. Wash slides in 70 % ethanol, 10 min.
5. Wash slides in 50 % ethanol, 10 min.
6. Rinse slides in distilled water. From this point onwards do not allow the slides to dry, as this might result in high background signal.
7. Cover sections with 20 μg/ml proteinase K in 0.1 M Tris–HCl pH 7.5 and 0.05 M EDTA for 10–20 min at 37 °C. This step might require optimization in terms of time of incubation and enzyme concentration. Under-digestion might lead to a lower specific signal, while over-digestion might lead to high background and poor tissue morphology.
8. Rinse slide 5× in distilled water.
9. Incubate slides in ice-cold 20 % (v/v) acetic acid for 20 s to permeabilize cells.
10. Dehydrate the sections for approximately 1 min each in 70 % EtOH, 95 % EtOH, and 100 % EtOH then air dry.
11. Mark around the tissue section with a PAP Pen to create a hydrophobic barrier. Add 100 μl hybridization solution to each section. Volume might vary pending the original area of the tissue section. Pre-hybridize the slides for 1 h in a humidified hybridization chamber (plastic dish with wet paper towels would suffice) at the desired hybridization temperature. While pre-hybridizing, dilute one or two probes (DIG-labeled RNA probe and Fluorescein-labeled RNA probe) in hybridization solution, and heat the probes for 95 °C for 2 min on a heat block. Chill on ice immediately to prevent rehybridization. Drain off the hybridization solution. Add 50–100 μl of diluted probe per section (ensure the entire sample is covered). Incubate in the hybridization chamber at 65 °C overnight.
12. Wash 3× 5 min at 37–45 °C with 50 % formamide/2× SSC.
13. Wash 15 min at 37–45 °C with 2× SSC.
14. Wash 30 min at 37–45 °C with 0.2× SSC.
15. Wash twice in MABT for 30 min at room temperature.
16. Incubate with superblock blocking buffer for 30 min at room temperature.
17. Add the anti-fluorescein-POD antibody (3) at the appropriate dilution in superblock blocking buffer. Incubate for 1–2 h at room temperature or 4 °C overnight.

18. Wash slides with MABT, 5× 10 min at room temperature. Wash twice for 5 min each in PBS.
19. Add TSA Plus Fluorescein Solution and incubate at room temperature for 5–30 min. Reaction time must be determined empirically for each probe (4).
20. Wash with 30 %, 50 %, 75 %, and 100 % methanol in PBS for 10 min each.
21. Incubate with 3 % H_2O_2 in PBS for 10–30 min to inactivate the first peroxidase (5).
22. Wash three times with PBS for 10 min.
23. Incubate with superblock blocking buffer for 30 min at room temperature.
24. Add the anti-DIG-POD antibody at the appropriate dilution in superblock blocking buffer. Incubate for 1–2 h at room temperature or 4 °C overnight.
25. Wash slides with MABT, 5× 10 min at room temperature. Wash 2× 5 min each in PBS.
26. Add TSA Plus Cy3 Solution and incubate at room temperature for 5–30 min. Reaction time must be determined empirically for each probe.
27. Wash twice with PBS for 10 min.
28. Cover tissue with Vectashield Mounting medium and place a coverslip of appropriate size and thickness.
29. Seal coverslip by covering its edges with nail polish.
30. Examine slides immediately or store at 4 °C in the dark.

4 Notes

1. The choice of size and sequence of the RNA probe is very important for a successful *in situ* hybridization. We introduce T7 promoter sequences to the 5¢ terminal of primers, which allows us to transcribe sense and antisense probes from a PCR fragment. Cloning a cDNA or a PCR product into a vector with a T7 promoter can substitute a PCR template for in vitro transcription.
2. In vitro transcription can fail for various reasons. One step we found to be important is to make sure that the enzyme buffer is fully reconstituted after thawing. For best results, role the tube fast between your gloved hands for about 10–15 s.
3. Although some protocols recommend removal of the DNA template with a DNase treatment prior to hybridization, we find that it is unnecessary since the amount of ssRNA molecules is significantly higher than the template dsDNA. Furthermore, DNase treatment and subsequent RNA purification often result in a significantly lower probe yields.

4. A fast way to know that your transcription was successful is to run 1 μl of your transcription reaction on a 1 % agarose gel. You should see a somewhat fuzzy but very dominant band at roughly the size of your DNA PCR template. If the band looks smeared, it is better to repeat rather than continue with the protocol.
5. We find that riboprobes labeled with DIG give a stronger signal than probes labeled with fluorescein. Consequently, we typically use the DIG-labeled probe for transcripts that are expressed at lower levels and fluorescein-labeled probes for more abundant transcripts.
6. For double immunostaining with primary antibodies from a single host, we use a relatively high concentration of conjugated Fab. We use Alexa Fluor 488-conjugated AffiniPure Fab antibody to saturate the first primary antigen. Since Fab is monovalent antibody, saturation of the first primary antibody blocks any subsequent binding of any additional antibodies.
7. Our protocol often results in high background of the first antigen tested, which often requires careful optimization. To increase signal-to-noise ratio, we typically lower the concentration of conjugated Fab, which is then followed by a saturation step of the sample by using excess of unconjugated Fab.

Acknowledgments

We thank Larisa Belaygorod, Phillip Karp, and Lynda Ostegaard for technical assistance and reagents. This work was supported by grants R03DC010244 from the NIH and MD-II-2009-170 from the Children Discovery Institute.

References

1. Buck L, Axel R (1991) A novel multigene family may encode odorant receptors: a molecular basis for odor recognition. Cell 65:175–187
2. Godfrey PA, Malnic B, Buck LB (2004) The mouse olfactory receptor gene family. Proc Natl Acad Sci U S A 101:2156–2161
3. Malnic B, Godfrey PA, Buck LB (2004) The human olfactory receptor gene family. Proc Natl Acad Sci U S A 101:2584–2589
4. Zhang X, Firestein S (2002) The olfactory receptor gene superfamily of the mouse. Nat Neurosci 5:124–133
5. Zozulya S, Echeverri F, Nguyen T (2001) The human olfactory receptor repertoire. Genome Biol 2:research0018.1–0018.12
6. Keller A et al (2007) Genetic variation in a human odorant receptor alters odour perception. Nature 449:468–472
7. Spehr M et al (2003) Identification of a testicular odorant receptor mediating human sperm chemotaxis. Science 299:2054–2058
8. Braun T et al (2007) Enterochromaffin cells of the human gut: sensors for spices and odorants. Gastroenterology 132:1890–1901
9. Shah AS et al (2009) Motile cilia of human airway epithelia are chemosensory. Science 325: 1131–1134
10. Karp PH et al (2002) An in vitro model of differentiated human airway epithelia. Methods for establishing primary cultures. Methods Mol Biol 188:115–137

Part IV

Functional Analysis of Olfactory Receptors

Chapter 13

Functional Characterization of Insect Olfactory Receptor Neurons Through In Vivo Approaches

Michel Renou and Philippe Lucas

Abstract

Insects rely significantly on olfactory cues for recognition and finding of vital resources such as food and mates. Odor detection is mediated by primary sensory neurons housed in individualized cuticular structures, the sensilla. Using microelectrode-based techniques, it is possible to record the nervous activity inside specific sensilla after odor stimulation. We describe in this chapter the methods routinely used in our laboratory to stimulate, record, and analyze the activity of olfactory receptor neurons in moths or flies.

Key words Insect, Olfactory receptor neuron, Electroantennogram, Antenna, Sensillum, Electrophysiology, Single sensillum recording, Action potential

1 Introduction

The main olfactory organ of insects is a pair of external appendages, the antennae. Insect antennae bear minute cuticular structures, the sensilla, which house the olfactory receptor neurons (ORNs, also designated as olfactory sensory neurons or OSNs). The sensilla are morpho-functional units that protect ORNs and create the environment that enables their function. There are different morphological types of olfactory sensilla, each with very different external shapes: plates (sensilla placodea), long hairs (sensilla trichodea), cavities (sensilla coeloconica), and pegs (sensilla basiconica). All possess wall pores that enable the entry of the odor molecules (1). This organization makes it possible to record the electrical activity from ORNs, extracellularly. ORNs can be chosen by their location in morphological types of sensilla or their defined places on the antennae.

Admittedly, most insect ORNs express a single type of olfactory receptor (OR) in their dendritic membrane. Thus, the response spectra of ORNs reflect the selectivity of ORs and their tuning. However, the sensillar lymph that bathes ORNs contains odorant binding proteins and odor degrading enzymes, whose contribution

Chiquito J. Crasto (ed.), *Olfactory Receptors: Methods and Protocols*, Methods in Molecular Biology, vol. 1003,
DOI 10.1007/978-1-62703-377-0_13,

to the selectivity and the kinetics of the response is critical, although not fully understood. Currently, the integrity of the entire system that insures the high sensitivity and specificity of the insect olfactory system can only be maintained in vivo. Thus, while in vitro approaches of OR function from various heterologous systems have been developed, in vivo studies are still necessary to fully understand the capacity of insect olfactory system to code the quality, intensity, and temporality of the olfactory signal.

ORNs are bipolar neurons that influence the transduction of the chemical signal into a neural code transmitted to the brain by the axons. Their activity results in two main electrical signals. First, under odor stimulation, the activation of ORs leads to a depolarization of the ORN, the receptor potential; this depolarization can be recorded locally as a change in the transepithelium potential (TEP), or globally, as an electroantennogram (EAG). Next, this odor-induced depolarization opens voltage-sensitive ion channels, leading to the generation of action potentials (spikes). All these electrical events may be recorded extracellularly by appropriate microelectrode techniques that are described hereafter.

2 Materials

The electrophysiological set-up includes ingredients that are commercially available from different sources. Therefore, we will indicate here brand names just as examples. We will include however a list, with examples of procedures designed to fulfill specific objectives or according to biological models. Building a setup requires addressing and finding solutions to several issues that are not necessarily global or generic. Challenges in obtaining elements used in our procedures might arise because of the lack of local facilities. Indeed, access to a properly equipped workshop can offer cheaper and better adapted alternatives. A strong constraint is to find a compromise between ergonomics and lack of space (Fig. 1).

2.1 Insect Preparation

1. Cylinder of CO_2.
2. Adhesive tape and double stick tape.
3. Insect minute pins.
4. Styrofoam.
5. Dental wax (e.g., Surgident Periphery Wax, Heraeus).
6. Drummond forceps.
7. Ophthalmological scissors.

2.2 Microelectrodes

The choice of extracellular microelectrodes depends on the objectives of the study and the shape and size of the sensillum housing the ORNs. Two main techniques are used: base recording with

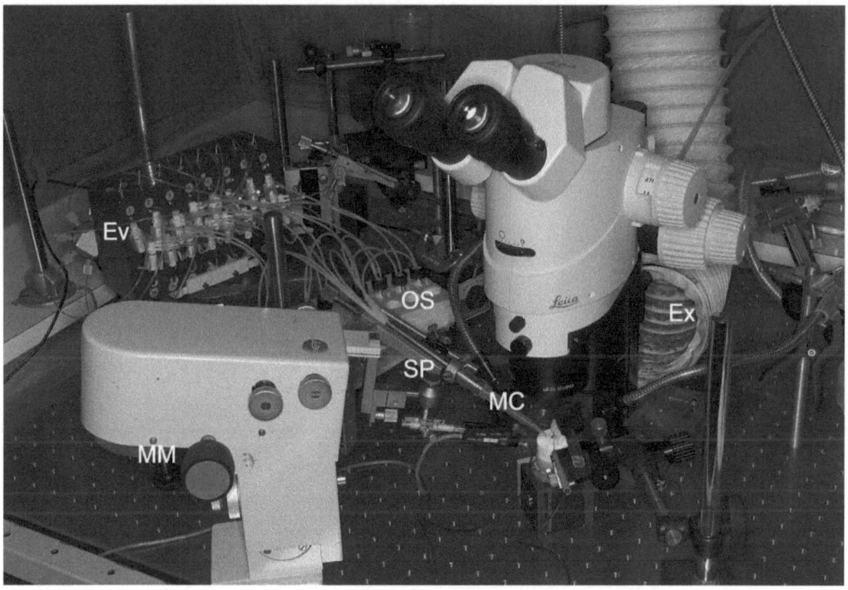

Fig. 1 Overview of the electrophysiological set-up equipped with a multichannel odor delivering device. The recording electrode is positioned at the base of one sensillum with a micromanipulator (MM). Electrovalves (Ev) control air flows through vials containing odor sources (OS). Odorized air flows converge to a multichannel stimulation pencil (SP). A 1-ml pipette tip serves as a mixing chamber (MC). Air is evacuated out of the room by an exhaust fan (Ex)

tungsten microelectrodes for short hairs and plate sensilla or tip recording with glass microelectrodes for the long trichoid sensilla (2). Recordings with tungsten microelectrodes are more stable, but glass microelectrodes are suitable for DC recordings because they do not polarize and allow injecting dyes or pharmacological agents or modify the ionic environment of ORNs.

2.2.1 Tungsten Microelectrodes

1. Stereomicroscope (100 × magnification) set in horizontal position and illumination device.
2. Tungsten rods (diameter: 130 μm, e.g., TW5-6 from Science Products) cut into 25–30 mm segments.
3. An electrode holder made of a hypodermic needle soldered to a male electrical pin (home made or, e.g., NL 04 from Neurolog).
4. Solution for the electrolysis: $NaNO_2$ 71 g, KOH 34 g in 100 ml of distilled water.
5. Variable AC power supply 0–14 V.
6. Transparent parallipipedic 4 ml tank (e.g., spectrophotometry quartz cell) on a Lab Jack.
7. Carbon rod (e.g., the electrode of a 4.5 V flat dry battery).

Table 1
Composition of solutions (in mM)

	Sensillum saline	Hemolymph saline	*Drosophila* EAG saline
KCl	172	6.4	5
Glucose	22.5	340	
HEPES	10	10	10
$MgCl_2$	3	12	4
$CaCl_2$	1	1	1
NaCl	25	12	120
Osmotic pressure (mOsm/L) (mannitol)	425	450	
pH (KOH)	6.5	6.5	7.2

8. Electrical wires to connect the carbon rod and the electrode holder to the AC power supply.
9. Manual manipulator (1 axis).

2.2.2 Glass Microelectrodes

1. Electrode puller (e.g., Sutter Flaming Brown P97 or Narishige PC-10).
2. Borosilicate glass capillaries with internal filament (outer diameter: 1.5 mm, internal diameter 0.86 mm, length 100 mm, e.g., Harvard Apparatus).
3. Electrode holder (silver wire soldered to an electrical male pin, or purchased, e.g., the E and Q series Warner Instruments).
4. Saline solutions. Initially designed by Kaissling and Thorson (3) based on the ionic composition of the sensillum lymph and the hemolymph of . We have modified their composition (4) (Table 1). Saline solutions may be prepared in advance and stored at –20° in aliquots (2 ml) for one experimental day.
5. 1 ml syringes fitted with a heat-elongated 200 ml pipette tip.

2.3 Micromanipulators, Optics, and Grounding

1. An anti-vibration table (e.g., TMC 63-511, 63×90 cm, Technical Manufacturing Corporation). A cheap solution is a heavy and flat iron plate positioned on tennis balls on a stable lab bench.
2. Grounding. A Faraday cage can be easily and cheaply built using a metallic mosquito net attached to a metallic structure. It should rest on its feet so as to be mechanically independent of the anti-vibration table. The Faraday cage, the anti-vibration

table, the microscope, and micromanipulators must be connected to a central grounding point (e.g., brass bar drilled and tapped for banana plugs) to eliminate extraneous electrical interferences. The central grounding point must be connected to the case ground of the amplifier. It is preferable to connect all the apparatus to the same power line. The noise level can be tested by connecting a signal generator (e.g., Signal stimulator from Syntech) or a 10 MΩ resistance between the electrodes.

3. Stereomicroscope (e.g., Leica MZ16 with 25×/9.5B oculars and 16× zoom) (see Note 2).
4. Illumination (e.g., Intralux 5100 light source with incident gooseneck light guide from Warner Instruments). The source of light must be outside the Faraday cage.
5. A precise micromanipulator is required to position the recording electrode (we are familiar with the mechanical Leitz-Leica micromanipulator with joystick, but there are many excellent alternatives, e.g., Narishige NMN-25 or the motor driven DC 3K with a PM 10 Piezo Translator from Märzhäuser).
6. Three coarse micromanipulators on magnetic stands (e.g., Narishige UM-3C) to hold the reference electrode, the odor delivery device, and sharpened forceps used to cut sensillar tips.
7. Stainless steel forceps for cutting the sensilla (e.g., Dumont #5, Fine Science Tools) fixed to a rod to be mounted on a micromanipulator.
8. Instrument sharpening stones (e.g., Arkansas oil stones from Fine Science Tools).

2.4 Stimulus Delivery Device

1. Source of compressed air (from building or cylinder).
2. Pressure regulator (e.g., Air Liquide).
3. Charcoal filter (e.g., Hydrocarbon traps Cat 22013, Restek) and particle filter (25 μm).
4. Humidifier: 250 ml gas washing bottle with fritted disk, filled with distilled water.
5. Flow meters with precision valves (e.g., floating ball flow indicators ShoRate, Brooks Instruments or digital flow meter PFM7 series, SMC Corporation).
6. Flow Y-divider and 5-port manifold (e.g., Upchurch P514 and P115, respectively).
7. PTFE tubing: 2 × 30 cm of 1/4″ mm (main flows), 8 × 20 cm of 1/8″ tube (between manifold and valves).
8. Plastic precision connectors (kits containing the diameters corresponding exactly to the tubing are available from, e.g., Upchurch).

9. 4 ml glass vials stoppered by Teflon-coated septum corks.
10. Hypodermic needles (18 g size, 40 mm length) that have been cut to remove the Luer-Lock connector.
11. A multichannel stimulation pencil, made by pasting together eight segments of 30 cm of PTFE 1/8″ tubes in a stainless steel pipe (8 mm internal diameter, 12 cm long). The PTFE tubes are cut to the brim with the pipe after pasting. A pipette tip is fixed at the down-flow extremity to serve as a mixing chamber.
12. 3-way microvalves (e.g., LHDA1233115H, The Lee Company).
13. Multichannel time programmer (e.g., ValveBank, Automate Scientific).
14. Air exhaust with air intake behind the preparation.

A more detailed description of the 8-channel stimulus delivery device is described elsewhere (5). *For a simple single-channel stimulus delivery system*, see Note 1.

2.5 Recording and Saving the Data

1. High impedance DC amplifier (10^{11}–10^{13} Ω; e.g., EX-1 + head-stage 4002 from Dagan or a Neurolog with NL102 pre-amplifier from Digitimer).
2. Audio output (connected to the amplifier output).
3. Acquisition board (e.g., DT9800 from Data Translation). Models vary according to the location of the board (inside the computer or connected to a USB port), resolution (minimum of 12 bits), number of analog input channels (at least two, one for the biological signal and one for the stimulus), sampling frequency (at least 50 kHz), input range ± 10 V, and it should enable external trigger.
4. Shielded BNC cables.
5. Synchronization cable between stimulator and acquisition board.
6. Software for piloting the acquisition board (we use a home-developed program Awave (6); commercially available software generally depend on the hardware, Data Translation acquisition boards for instance may be piloted through Measure Foundry) (see Note 3).
7. Microcomputer station with two hard disks (one for programs and the other for saving data).

3 Methods

3.1 Preparing Stimulus Source

1. Under a hood, prepare dilutions of odors in an appropriate solvent, typically hexane (CAS 110-54-3) or mineral oil (CAS 8042-47-5) for highly volatile compounds.

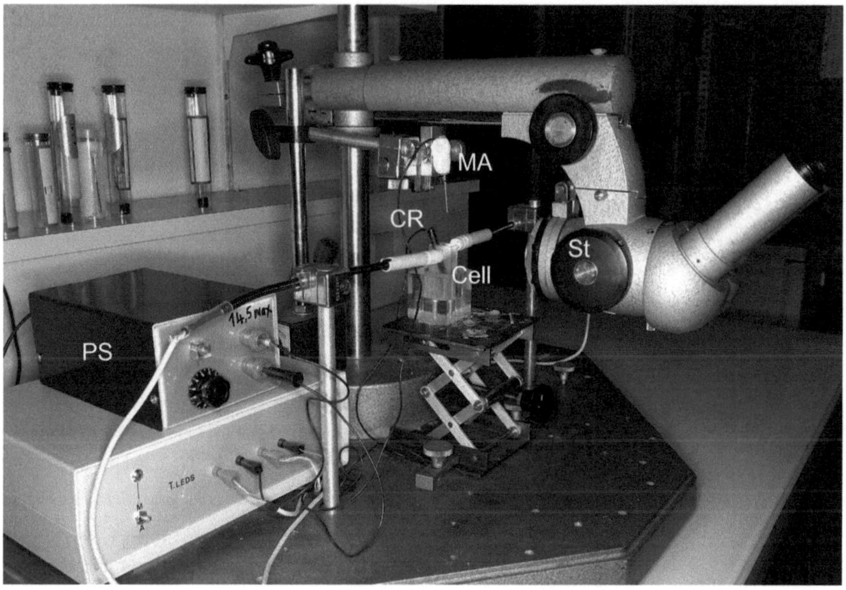

Fig. 2 Overview of a tungsten electrode etcher. A single axis manipulator (Ma) is used to dip back and forth the tip of an electrode into a quartz cell (Cell) that contains the electrolyte and a carbon rod (CR). The electrode holder (hypodermic needle) and carbon rod are connected to an AC power supply (PS). Progress of electrolysis is checked through a stereomicroscope (St)

2. Keep stock solutions at –20 °C to prevent solvent evaporation and compound degradation. New odor sources should be prepared daily.
3. To prepare stimulus source from hexane solution, pipette the needed volume of solution on to a piece of filter paper (e.g., Whatman paper), let the solvent evaporate, and insert the paper with clean forceps into a 4-ml vial.
4. To prepare stimulus sources from mineral oil solutions, pipette 1 ml of mineral oil into the vial, complete with the appropriate volume of compound (10 μl for a 1 % dilution v/v), close the vial with the PTFE septum, screw the cork, and vortex. Insert the needles through the septum and fix the vial in the stimulator.
5. Connect the stimulation pencil to the source vials by inserting a hypodermic needle in each PTFE tube (Fig. 1).

3.2 Preparing Electrodes

3.2.1 Tungsten Microelectrodes

1. Bend the tungsten wire at one extremity and introduce it into the electrode holder.
2. Fix the electrode holder to the manipulator. Set the AC power supply to ~8 V (Fig. 2).
3. Controlling the progress of the electrolysis through the microscope, repeatedly dip and withdraw the extremity of the wire into the solution. To produce a sharp but strong tip, it is better

to thin a ~0.5–1 mm of the wire first, and then complete the sharpening of the tip. The voltage of the source may be reduced for the final etching.

4. Rinse the electrode in distilled water. Several electrodes can be prepared in advance and stored in a dust-proof box. Optimizing electrode sharpening requires practice.

3.2.2 Glass Microelectrodes

1. Silver wires that serve as both electrodes are chlorinated by immersing them in concentrated bleach solution for 10–20 min until the silver is covered with a homogenous black coating layer (until the wire does not shine anymore). This prevents electrodes from polarizing and the potential from drifting with time. This process has to be repeated whenever the baseline drifts during experiments or when the offset between electrodes becomes too large to be compensated by the amplifier.
2. Fire-polish the extremities of glass capillaries to prevent scratching the chlorinated silver wire with electrodes.
3. Insert the glass capillary into the electrode puller.
4. Set the heating temperature and pulling strength to obtain the desirable tip. After pulling, glass microelectrodes are stored in a dust-proof box.
5. Break the tip of the recording electrode at the desirable diameter (~10 μm) under the microscope. Alternatively, electrodes can be pulled with a patch clamp puller (using a multi-step program) to make electrodes having the correct tip diameter.
6. Thaw a saline aliquot. Electrodes must be filled immediately before use. Fill the electrode using a syringe with an elongated 200 μl pipette tip. Bubbles can be dislodged by tapping the electrode with a finger. Before filling, the recording electrode tip may be dipped in a solution of 0.5 % polyvinylpyrrolidone in the same saline to limit evaporation and dialysis of the sensillar lymph (3).
7. Slip the glass electrode over the chlorinated silver wire connected to the electrode holder.

3.3 Preparing the Insect

Insect movements cause artifacts in the recordings or a loss of contact. The insect must be carefully immobilized in a system of contention adapted to its size. The insect may be anesthetized with CO_2 to facilitate its handling.

3.3.1 Moths

1. Immobilize the moth in a Styrofoam block with the head protruding.
2. Fix strips of adhesive tape around the neck to further restrain it.

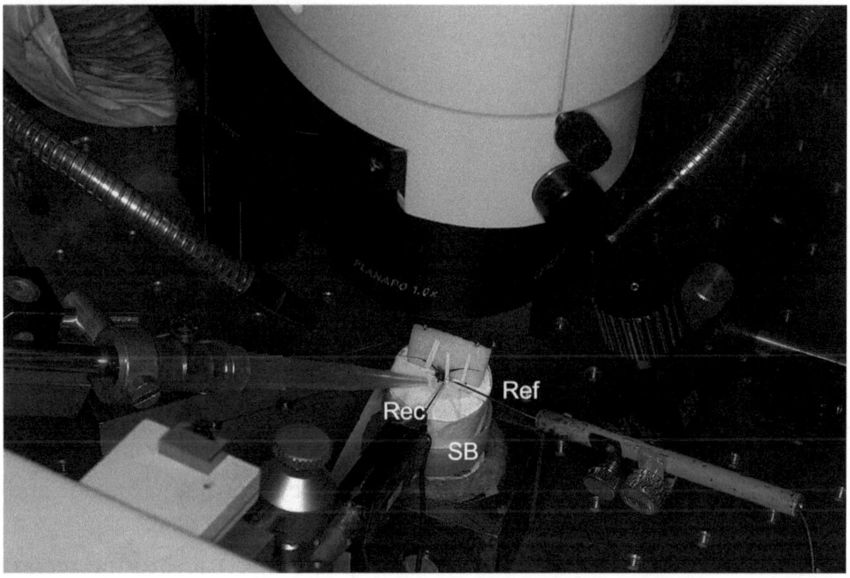

Fig. 3 Preparation for recording the activity of moth ORNs with tungsten electrodes. The moth is immobilized in a Styrofoam block, with the head protruding and antennae restrained with stripes of tape. The reference electrode (Ref) is inserted into the eye and the reference electrode (Rec) is implanted at the base of a sensillum. The output of the stimulus delivering device is focused on the antenna (St)

3. For SSR recordings, attach one antenna with two stripes of tape to a Styrofoam surface covered with a double stick tape (Fig. 3).

3.3.2 Drosophila Flies

1. Cut a 200-µl pipette tip at its narrow end at a diameter slightly smaller than the diameter of the fly head.
2. Under a stereomicroscope, insert the fly in the pipette tip with its head protruding.
3. Immobilize it with dental wax. Only part of the head should protrude from the tip, giving access to antennae but preventing the fly head moving freely.

3.4 Recording

3.4.1 EAG Recording

Antennae have different shapes: filiform (lepidoptera), lamellate (scarab beetles), or globular (diptera). Moreover, distribution of sensilla with specific ORNs may be heterogeneous. This impacts the way the microelectrodes should be used to record EAGs.

For a moth:

1. Insert a chlorinated silver wire into the abdomen.
2. Connect this wire to the ground entry on the headstage to serve as the reference electrode. An alternative is to use a glass electrode connected to the ground via a silver wire and filled

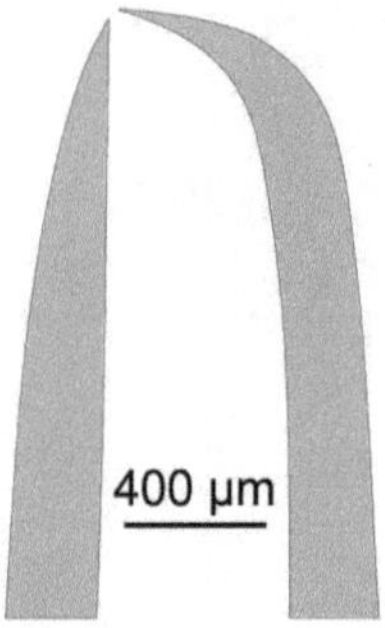

Fig. 4 Shape of branch tips of the modified forceps used for cutting the tip of trichoid sensilla

with the same solution as the recording electrode and to insert it anywhere in the insect body (e.g., neck).

3. Cut out the antenna tip and insert it into the recording electrode filled with hemolymph saline.

For *Drosophila*:

1. Insert two glass electrodes with a tip diameter of 1–3 µm in the electrode holders.
2. Implant the reference electrode in the eye, not too deep to prevent dehydration of the preparation.
3. Press the recording electrode on the third antennal segment until the signal baseline is thin and stable to obtain a good signal-to-noise ratio.

3.4.2 SSR Tip Recording

1. Prepare the sensillum tip cutting device by fixing a forceps on a metal rod.
2. Bend one branch of the forceps to make scissor-like tips.
3. Sharpen the extremities on the oil stone, finish polishing with another forceps. At the end the bent branch should very slightly scrap the other branch when pressing the forceps (Fig. 4).
4. Focus the light guides onto the antenna (see Note 4).
5. Bring modified forceps in close vicinity to the sensilla with a micromanipulator.
6. Cut the tip of one or several sensilla by pressing gently the forceps. Cut sensilla remain functional for a long time (≥2 h).
7. Insert the reference electrode into the adjacent segment to minimize contributions of field potentials (EAG signal).
8. Under the stereomicroscope, briefly dip the tip of the recording electrode into a drop of mineral oil to prevent drying, which may cause an increase in salt concentration, resulting in crystallization.

9. At high magnification, adjust the focus on the desired sensillum tip and bring the tip of the microelectrode in contact with it. Movements in the Z direction are judged by adjusting the microscope focus. A successful contact with the sensillum will appear in the form of spontaneous firing of action potentials. It is thus convenient to monitor the electrical activity on the audio output while looking at the preparation through the microscope.
10. Check the signal-to-noise ratio on computer screen or oscilloscope and move the electrode to another sensillum if it is not good enough. A good quality of recording results in easy and fast spike sorting.

3.4.3 SSR Tungsten Recording

1. Insert the reference electrode anywhere in the insect body (e.g., antenna for moths, eye for *Drosophila*).
2. Approach the recording electrode to the antennae using the coarse movement of the micromanipulator, taking care not to damage the tip of the electrode as it is very delicate.
3. Under high magnification of the microscope and with the fine micromanipulator movement bring the electrode close to the aimed sensillum (Fig. 3).
4. Gently push the electrode against the antenna until it pierces the cuticle at the base of the sensillum, with an angle of penetration of ~90 °.
5. Monitor the firing activity on the audio output. Depending on the species, the depth of electrode penetration may have to be adjusted according to the signal-to-noise ratio of the recorded activity. Going too deep may result in a strong a transient firing activity corresponding to ORN depolarization and death. Once a stable signal is recorded and if the preparation does not move, the activity of a sensillum can be recorded for several minutes. Continuous contact to a sensillum can be made without changing the tungsten electrode as long as its tip is not bent.

3.5 Stimulation Program and Recording

1. Open the source of air. Adjust the main air flow with the flow meter to ca 1,800 ml/min to obtain a flow rate of 220 ml/min for each channel after division of flows.
2. Focus the output of the stimulus delivering device onto the antenna; keep a constant distance between antenna and output for all preparations (Fig. 3).
3. Adjust the gain on the amplifier to ×100–1,000 and low pass filter to 5 kHz.
4. Use the input-offset control to cancel potentials in DC mode (EAG and sensillar potential recordings).
5. Adjust the high pass filter to 100 Hz for tungsten recordings or for tip recordings when only the firing activity is studied.

6. Send a control stimulus to check for the absence of response to pure air. Unexpected response to a blank stimulus—pure air or solvent—reveals a contamination.
7. Start the stimulation sequence. Stimulation protocols should avoid long odor presentation (>0.5 s) unless required by the experiment, manage large interstimulus time (from 30 s up to several minutes at very high odor concentration), and randomize the order of presentation of the stimuli (see Note 5). However, for dose–response studies use increasing doses are preferred to random presentation.

3.6 Analyzing the Activity of ORNs

Digitized recording are stored in the computer as data files containing the electrical signal continuously recorded for several seconds from a sensillum, event marks, and information about the recording parameters and stimulation conditions. Stimulations are generally short (between 0.1 and 5 s) so that saving 60 s of recording enables to consider the pre-stimulus as well as the post-stimulus activity and to analyze not only the intensity of the response but also its dynamics. Considering the storage capacity of hard disks nowadays, it represents low volume (2.3 Mo per minute of recording at 10 kHz sampling rate).

1. Extract spikes from the noise; algorithms for spike detection generally consider short events crossing a threshold.
2. Sort spikes emanating from the different ORNs housed in a single sensillum into classes (spikes can usually be sorted by their amplitude).
3. Export spike occurrence times in xls, csv, or txt format to analyze the firing by custom-developed programs.
4. An efficient and simple quantification of the ORN response to a given stimulus is typically made by counting the number of spikes fired in a given time window (for instance 0.5 s) beginning at the stimulus onset and subtracting the number of spikes fired before stimulation and it is expressed in spikes/s. Stimulus onset is often marked as the opening of the valve, but the effective time for arrival of odor on the antenna is delayed (in the range of 0.2 s) and varies according to the length of tubing between the valve and the odor delivery device, air speed in this tubing and distance of the antenna from the odor delivery. Thus, using a corrected stimulus onset time by adding this time latency add rigor to the processing.
5. Observe the temporal dynamics of spike trains by for instance, building peri-stimulus time histograms (PSTH) from counts of spikes by short time bins (for instance, 50 ms). Some treatments or physiological conditions can affect the latency of the response or the decay time which can be computed from

PSTH. Firing frequency measures and automatic detection of responses can be achieved by algorithms as described by Blejec (7).

4 Notes

1. Multichannel stimulus delivery devices are not commercially available ready to use. Alternatively, simpler but very reliable single channel stimulus delivery systems can be used (Fig. 5). A ready-to-use stimulus controller device may be purchased from Syntech (CS-55).
2. A stereomicroscope offers a maximum magnification (×300) largely sufficient for tip recording. Single sensillum recording with tungsten microelectrodes from small preparations (e.g., *Drosophila*, mosquitoes) requires an upright microscope (e.g., Olympus BX51 WI with 10× oculars and 5× and 50× air objectives).
3. Developing an application to pilot the acquisition board requires programming expertise. Manufacturers of electrophysiology equipment propose integrated software, enabling recording and analysis of nervous activity. The software is

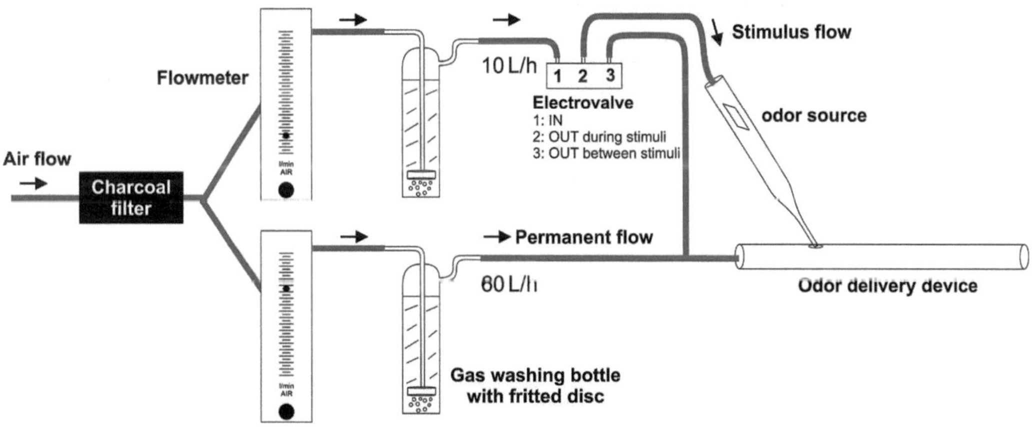

Fig. 5 A single channel odor delivery system using a Pasteur pipette as pheromone source. The odor delivery device is a 20 cm long tube, 7 mm inner diameter, with a hole to insert the tip of a Pasteur pipette at 15 cm from the outlet of this tube. A constant charcoal-filtered and humidified airflow is delivered on the antenna at a flow of 60 l/h. Stimuli are applied by inserting a Pasteur pipette containing a filter paper impregnated with an odor source in the glass tube. The activation of the electrovalve blows an air pulse of 10 l/h (#2: OUT during stimuli) through the Pasteur pipette during a controlled duration. The odorized air flow is carried away by the permanent main flow of humidified air. Flow rates are kept constant, and therefore the mechanical component of the stimulation is minimized, by connecting the second outlet of the electrovalve (#3: OUT between stimuli) to the continuous airflow

hardware-dependent and must be purchased together with the acquisition board. Such programs include functions for the analysis of firing activity (e.g., Autospike from Syntech, pClamp from Molecular Devices).

4. Color fringes sometimes make viewing single hairs difficult. In which case, illumination from behind the antenna improves visibility.
5. Stability of the preparation in time is critical. Some preparations live shorter than the necessary time for presenting all stimuli. A control stimulus may be used periodically to check the stability and compensate for a potential gradual loss of sensitivity. In some species, it is possible to record from an isolated antennae, but the life span of the preparation is then reduced. Sensory fatigue or adaptation can be prevented by using lower concentrations of odorants and increasing the interstimulus interval.

References

1. Steinbrecht RA (1996) Structure and function of insect olfactory sensilla. Ciba Found Symp 200:158–177
2. Kaissling KE (1974) Sensory transduction in insect olfactory receptors. In: Jaenicke L (ed) Bio chemistry of sensory functions. Springer, Berlin
3. Kaissling KE, Thorson J (1980) Insect olfactory sensilla: structural, chemical and electrical aspects of the functional organization. In: Satelle DB, Hall LM, Hildebrand JG (eds) Receptors for neurotransmitters, hormones and pheromones in insects. Elsevier-North Holland, Amsterdam
4. Pezier A, Acquistapace A, Renou M, Rospars JP, Lucas P (2007) Ca2+ stabilizes the membrane potential of moth olfactory receptor neurons at rest and is essential for their fast repolarization. Chemical Senses 32:305–317
5. Party V, Hanot C, Said I, Rochat D, Renou M (2009) Plant terpenes affect intensity and temporal parameters of pheromone detection in a moth. Chemical Senses 34:763–774
6. Marion-Poll F (1995) Object-oriented approach to fast display of electrophysiological data under MS-Windows. J Neurosci Methods 63:591–592
7. Blejec A (2005) Statistical method for detection of firing rate changes in spontaneously active neurons. Neurocomputing 65–66:557–563

Chapter 14

Functional Assay of Mammalian and Insect Olfactory Receptors Using *Xenopus* Oocytes

Charles W. Luetje, Andrew S. Nichols, Ana Castro, and Benjamin L. Sherman

Abstract

The large number of olfactory receptors (ORs) expressed by various mammalian and insect species, as well as the large number of potential odorant ligands, has made the pairing of odorants with receptors (de-orphaning) exceedingly difficult. These efforts are further complicated by difficulties in expressing ORs in many standard expression systems. *Xenopus laevis* oocytes offer a versatile expression platform for the de-orphaning and functional characterization of ORs. Two-electrode voltage clamp electrophysiology is a common and relatively straightforward approach to the functional assay of receptors expressed in *Xenopus* oocytes, and this technique has been discussed extensively in the literature. However, laboratories that are new to the use of *Xenopus* oocytes are often stymied by some of the peculiarities of the *Xenopus* oocyte expression system. We discuss some of the key methodological issues in *Xenopus* care, oocyte isolation and receptor expression, with a focus on using this expression system to study the ORs of mammals and insects.

Key words *Xenopus* oocyte, Electrophysiology, Olfactory receptor, G-protein-coupled receptor, Ligand-gated ion channel

1 Introduction

The functional analysis of olfactory receptors (ORs) of mammals and insects presents a number of challenges to the investigator. In particular, the large number of ORs in each species, as well as the large number of potential odorant ligands, has made pairing of odorants with receptors (known as "de-orphaning") difficult (1). In addition, difficulty in achieving expression has hampered efforts to study these receptors (2, 3).

The oocytes of the African Clawed Frog (*Xenopus laevis*) have been used for decades as a functional expression system for a wide variety of receptors and channels. This system is particularly useful for the expression and characterization of "difficult" receptors.

Chiquito J. Crasto (ed.), *Olfactory Receptors: Methods and Protocols*, Methods in Molecular Biology, vol. 1003,
DOI 10.1007/978-1-62703-377-0_14, © Springer Science+Business Media, LLC 2013

Xenopus oocytes can be used for pharmacological analyses, biophysical investigations, and receptor mutagenesis studies (4–8). Recently, high-throughput screening approaches have also become available (9). Thus, the utility of this system for the de-orphaning and functional characterization of ORs is clear.

The large size and sturdy constitution of *Xenopus* oocytes has made electrophysiology, generally by two-electrode voltage clamp, a preferred functional approach. Electrophysiology with oocytes is relatively straightforward and has been covered in detail elsewhere (10). However, laboratories that are new to the use of *Xenopus* oocytes are often stymied by some of the peculiarities of this expression system. Here we discuss some of the key methodological issues in *Xenopus* care, oocyte isolation and receptor expression, with a focus on using this expression system to study the ORs of mammals and insects.

2 Materials

2.1 Frog Care

2.1.1 Feeding and Health Issues

1. Adult, female *Xenopus laevis* frogs (Nasco, 9+ cm Mature Female).
2. "Frog Water": Sodium Bicarbonate buffered RO water (0.02 % w/v). ~125 gal per week.
3. Frog brittle (Nasco).
4. Gentamicin (Sigma).
5. Streptomycin (Sigma).
6. Methylene Blue (Kordon).
7. Water test kit (Hach).
8. Proline Freshwater Nitrifying Bacteria (Aquatic Eco-Systems).
9. The frogs must be housed in a continuously circulating aquarium system. We use an X-Rack system from Aquatic Habitats, Inc. An aquarium system can also be constructed and appropriate components are available from Aqua Logic, Inc. To properly maintain the frogs, the system must include continuous circulating water (with several complete turnovers per hour), particulate filtration, charcoal filtration, porous media to support a bacterial colony, in-line UV sterilization, and an in-line chilling system.

2.2 Surgical Removal of Oocytes

1. Anesthetic solution: 1.2 g 3-amino benzoic acid ethyl ester (Tricaine-S, Western Chemical), fully dissolved in 1.5 L Frog Water, pH adjusted to 7.0 with sodium bicarbonate.
2. Lab diapers.
3. PVP Iodine, OVADINE (Western Chemical).
4. Cellophane wrap.

5. Surgical Instruments: Moloney Forceps (Roboz), Micro Dissecting Scissor (Roboz), Dumont Forceps (Fine Science Tools), Kelly Hemostat (Fine Science Tools). Several sets are maintained. Each set is cleaned and autoclaved after surgery.
6. Barth's Ca-free solution (see below).
7. 4-0 Chromic gut suture (Roboz).
8. 1 mL syringe and needle (Becton Dickinson).
9. Gentamicin (Sigma).

2.3 RNA Preparation

2.3.1 Useful Plasmids

1. pGEMHE (11).
2. pCI (Promega).
3. pSP64, pSP65 (Promega).

2.3.2 DNA Linearization

1. Appropriate restriction enzyme.
2. UltraPure phenol:chloroform:isoamyl alcohol, 25:24:1, v/v (Invitrogen).
3. NaOAc 3 M, pH 5.2.
4. 100 % EtOH.
5. 70 % EtOH (made with DEPC H_2O).
6. DEPC H_2O (see Note 1). In a fume hood, 100 μL diethyl pyrocarbonate (Sigma) is added to 100 mL H_2O. The bottle is tightly capped, shaken vigorously and tapped to disrupt bubbles. The cap is loosened and the bottle is maintained in the fume hood overnight, then autoclaved.

2.3.3 RNA Synthesis

1. RNA synthesis kit (mMessage mMachine or mMessage mMachine ULTRA from Applied Biosystems).
2. 70 % EtOH (made with DEPC H_2O).
3. 100 % EtOH.
4. 10 M NH_4OAc (made with DEPC H_2O and sterile-filtered).
5. 1.5 mL low adhesion microcentrifuge tubes (USA Scientific).

2.4 Preparation of Oocytes

1. Barth's regular solution (88 mM NaCl, 1 mM KCl, 2.4 mM $NaHCO_3$, 0.3 mM $CaNO_3$, 0.41 mM $CaCl_2$, 0.82 mM $MgSO_4$, 15 mM HEPES, pH 7.5 and 100 μg/mL Ceftazidime): 40 mL Barth's "A" Stock, 40 mL Barth's "B" Stock, 100 mg Ceftazidime, QS 1 L H_2O. Sterile filtered with 0.45 μm bottle top filter (Corning). Stored in foil wrapped glass bottle at 4 °C.
2. Barth's Ca-free solution: 40 mL Barth's "A" Stock, 40 mL Barth's "B Ca free" stock, no antibiotic, QS 1 L H_2O. Sterile filtered with 0.45 μm bottle top filter (Corning). Stored in foil wrapped glass bottle at 4 °C.

3. Barth's "A" stock: 128 g NaCl, 2 g KCl, 5 g $NaHCO_3$, 89 g HEPES, pH adjusted to 7.6 with HCl or NaOH, QS 1 L H_2O. Stored in 40 mL aliquots at −20 °C.
4. Barth's "B" stock: 1.9 g $CaNO_3$•$4H_2O$, 1.5 g $CaCl_2$•$2H_2O$, 5 g $MgSO_4$•$7H_2O$, QS 1 L H_2O. Each reagent is completely dissolved before adding the next. Stored in 40 mL aliquots at 4 °C.
5. Barth's "B Ca free" stock: 5 g $MgSO_4$•$7H_2O$, QS 1 L H_2O. Stored in 40 mL aliquots at 4 °C.
6. 20 mL Collagenase B solution: 1.0 mg/mL Collagenase B (Roche 11088831001) in Barth's Ca-free solution (see Note 2).

2.5 Oocyte Injection

1. Mineral oil.
2. DEPC H_2O (Sigma).
3. 70 % EtOH (made with DEPC-treated H_2O).
4. Glass micropipettes (Drummond).
5. Barth's regular solution.

2.6 Electrophysiology in Oocytes

1. ND96 buffer (96 mM NaCl, 2 mM KCl, 1 mM $CaCl_2$, 1 mM $MgCl_2$, 5 mM HEPES, pH 7.5): 1,000 mL 200 mM HEPES, pH 7.4, 1,136 mL 2.5 M NaCl, 28 mL 3 M KCl, 40 mL 1 M $CaCl_2$, 40 mL 1 M $MgCl_2$, QS 40 L H_2O. Prepared in a large carboy. Stored at room temperature.

3 Methods

3.1 Frog Care

1. Proper care of the *Xenopus* colony is essential to the production of oocytes that will be suitable for receptor expression and functional experiments.
2. Adult, female *Xenopus laevis* frogs are maintained in an aquarium system at a density of no more than one frog per gallon.
3. Aquaria are maintained at 60–64 °C using an in-line chiller system.
4. The aquaria room is kept on a 12-h light/12-h dark cycle.
5. The system should continuously circulate water, with several complete turnovers per hour.
6. 10–15 % of system volume should be replaced with fresh Frog Water, 5 days a week.
7. Water quality is tested two times per week. Adjustments are made if the pH falls outside of 6.5–6.8. Any detectable ammonia or nitrite must be corrected immediately by taking the following steps. Replace 50 % of the system volume with

fresh Frog Water. Change the charcoal. Turn off the UV sterilizer and reinoculate the system with bacteria for the bacterial filter.

8. The crude particulate filter (>150 μm) pad is replaced daily. The fine particulate filter (>50 μm) cartridge is replaced monthly. The activated charcoal filter is emptied and refilled monthly. The bacterial colony filter is reinoculated yearly (59 mL/5 gal). The bulb for the in-line UV sterilizer is replaced yearly or as needed.
9. Frogs are fed three times per week (Monday, Wednesday, and Friday). Frogs are fed an amount of Frog Brittle that they can consume within 10 min. After 10 min all excess food is removed. Frogs must not be fed within 12 h of surgery.
10. "Red leg" disease, a common name for bacterial septicemia in frogs, is identified by cutaneous hemorrhages, particularly on the legs and feet. Red leg can be lethal. To reduce mortality rates, treatment must be initiated at the first indication of infection. The affected frog is euthanized. In addition, the entire aquaria system is treated with an antibiotic cocktail (10 mg/mL streptomycin and 10 mg/mL gentamicin). Aquaria circulation is turned off. Antibiotic cocktail is added, 5 mL for every 75 L of water (typical tank size). After 3 h, circulation is turned on. Treatment is conducted once.
11. Fungal infections are identified by white cotton-like spots on the skin (different presentations have been observed by others). Aquaria circulation is turned off. Methylene Blue is added, 10 mL for every 75 L of water (typical tank size). After 6 h, circulation is turned on. Twenty-four hours after treatment, a 30 % water change is conducted, and filters and charcoal are changed.

3.2 Surgical Removal of Oocytes

1. Fresh anesthetic solution is prepared for each surgery.
2. A frog is removed from general housing and placed in a 4 L lidded plastic chamber containing 1.5 L anesthetic solution for 15–30 min. An appropriate level of anesthesia is reached when the nasal flare and swallow reflexes are lost. Nasal flare occurs when the top of the head is gently touched between the nostrils. Swallow reflex occurs when the throat is gently rubbed. Appropriate level of anesthesia must not be exceeded as lethal overdose can occur.
3. The surgical station is prepared by thoroughly cleaning the bench with alcohol and positioning a new lab diaper, absorbent side up. The diaper is wetted with Frog Water.
4. After an appropriate level of anesthesia is achieved, the frog is removed from the anesthetic solution and positioned on the diaper with the ventral side up and the head pointing away from the investigator.

5. The abdominal area is cleaned with iodine. A folded piece of surgical gauze is held with forceps, soaked with iodine, and applied to the abdominal area with gentle circular motions. In the same fashion, a new piece of gauze is used to wipe off the excess. The application and wipe is repeated three times, using new gauze each time.
6. As an added precaution against contamination from the skin, cellophane wrap is applied to the abdomen.
7. The incision site is selected, in the lower left or lower right quadrant of the abdomen, about 1.5 cm lateral of midline. A 1 cm incision is made in the skin (and the cellophane wrap) using tweezers and scissors. A 1 cm incision is then made in the abdominal muscle layer using tweezers and scissors.
8. Using tweezers, two or three ovary lobes are pulled through the incision, laid on the cellophane wrap that is covering the skin and cut away. The ovary lobes are placed in a plastic petri dish containing Barth's Ca-free solution for subsequent oocyte preparation.
9. The incision is sutured. First, the muscle is sutured. Second, the skin is sutured. Typically, four to six stitches are required for each layer.
10. The incision site is treated with two subcutaneous injections totaling 100 μL Gentamicin (10 mg/mL), immediately above and below the incision site.
11. Any excess air or fluid (seen as a bubble under the skin) is removed through the sutured incision with an empty syringe and needle.
12. The anesthetic chamber is rinsed thoroughly. Wetted paper towels are placed inside the chamber to create a humid environment. The frog is placed in the chamber ventral side up and allowed to recover. The frog should be kept moist by periodic spraying with water. Recovery is determined by self-righting of the frog. The lid must be securely attached or weighed down, as *Xenopus* frogs can easily escape.
13. After recovery, the frog is returned to the general population.
14. Any suture remnants are removed at the time of the next surgery.
15. Surgery is performed on each frog a maximum of four times. Surgeries are at least 3 months apart and on alternating sides of the abdomen. After the fourth surgery, the animal is euthanized (3 h in anesthetic solution, followed by pithing).

3.3 RNA Preparation

3.3.1 Plasmid Constructs

1. Any plasmid with an RNA polymerase promoter (SP6, T3, T7) will work. We currently favor using pGEMHE (11) and pCI (Promega).
2. Successful functional expression of some mammalian ORs requires the presence of the "rho-tag." This consists of 20 amino acid residues from the amino-terminus of mammalian rhodopsin and is appended to the amino-terminal end of the OR. "Rho-tags" from bovine (12) and human (13) rhodopsin have been successful. We are currently using the human sequence (see Note 3).

3.3.2 DNA Linearization

1. A restriction enzyme should be chosen that allows linearization of the plasmid 3′ of the receptor open reading frame.
2. A 100 μL reaction is prepared: QS 100 μL H_2O, 10 μL 10× BSA (if required by enzyme), 10 μL 10× buffer, 10 μg DNA (up to 10 μL volume), 10 U/μg DNA of enzyme.
3. The reaction is incubated for 2 h in a 37 °C water bath.
4. 100 μL UltraPure phenol:chloroform:isoamyl alcohol is added to extract the DNA.
5. The reaction is vortexed for 1 min and spun down in a microcentrifuge (14,000 rpm, 18,000 × *g*) for 1 min.
6. The aqueous (top) phase is removed and precipitated by adding 10 μL NaOAc 3 M pH 5.2 and 250 μL cold 100 % EtOH.
7. The reaction is incubated for 1 h at −80 °C and then pelleted in a microcentrifuge (14,000 rpm, 18,000 × *g*) for 20 min.
8. The supernatant is aspirated and the pellet washed with 30 μL cold 70 % EtOH.
9. The pellet is spun down in a microcentrifuge (14,000 rpm, 18,000 × *g*) for 15 min.
10. The supernatant is aspirated and the pellet is air dried for 1 min.
11. The pellet is resuspended in 10 μL DEPC H_2O.

3.3.3 RNA Synthesis

1. RNA is synthesized according to the protocol supplied with the mMessage mMachine kit, using the lithium chloride precipitation method (see Note 4).
2. After addition of the lithium chloride, the reaction is incubated for 1 h at −20 °C.
3. Pellet the RNA in a refrigerated centrifuge (40,000 × *g* for 15 min).

4. The supernatant is carefully aspirated. The pellet is washed with 900 μL cold 70 % EtOH.
5. Re-pellet the RNA in a refrigerated centrifuge (40,000 × g for 10 min).
6. The EtOH is carefully aspirated.
7. The pellet is air dried for 1 min.
8. The pellet is resuspended in 47 μL DEPC H_2O and incubated for 2 min at in a 65 °C water bath.
9. 2 μL RNA is retained and run on a 0.8 % agarose gel to confirm that the RNA is intact.
10. 2 μL RNA is retained to measure the concentration using a spectrophotometer.
11. To the remaining 43 μL RNA, 10.75 μL NH_4OAc 10 M and 107.5 μL EtOH are added.
12. The RNA is stored at −80 °C.

3.4 Preparation of Oocytes

1. Oocytes are prepared immediately after frog surgery.
2. Working under Barth's Ca-free solution and using a sterile scissors, the lobes of ovary are cut into small pieces (~5–10 mm) and rinsed three to four times with Barth's Ca-free solution.
3. With a sterile transfer pipette, oocytes are aliquoted into 15 mL conical tubes, filling each tube with oocytes to the 2 mL mark. Excess solution is removed.
4. 10 mL Collagenase B solution is added. Tubes are incubated at room temperature on a rotisserie rotator (8 rpm) for 80 min.
5. Oocytes are rinsed with Barth's Ca-free solution and excess solution is removed. 10 mL fresh Collagenase B solution is added. Tubes are incubated on the rotator for an additional 80 min.
6. Oocytes are rinsed with Barth's regular solution several times until the solution is clear.
7. Oocytes are stored at 18 °C, submerged in Barth's regular solution in a petri dish.
8. Under a dissection microscope, Stage V–VI oocytes (approximately 800–1,200 μm in diameter) are selected for injection and placed in an autoclaved 5 cm glass petri dish with lid using a sterile plastic transfer pipette (see Note 5).
9. Oocytes can be injected with RNA anytime during the 24-h period after preparation.

3.5 Oocyte Injection and Care

1. An oocyte injection stage is constructed by affixing a 2 cm diameter piece of plastic mesh (Small Parts, Inc.) to a plastic tissue culture dish lid (Falcon 35 × 10 mm, Becton Dickinson)

using silicone caulk. The silicone should be at the outer edge of the mesh and should form a low rim to allow retention of solution on the mesh. The injection stage should be stored in 70 % EtOH and rinsed thoroughly with Barth's regular solution before use.

2. RNAs to be used for injection are aliquoted into low adhesion microcentrifuge tubes. To reduce the possibility of RNase contamination of an entire injection set, individual RNA aliquots are made to accommodate the injection of approximately 30–50 oocytes.
3. The RNA sample is pelleted in a microcentrifuge (14,000 rpm, 18,000 × *g*) for 15 min at room temperature. Pelleting occurs because the RNA was stored in a precipitated suspension. After centrifugation, the supernatant is immediately aspirated (see Note 6).
4. The RNA pellet is washed using 50 μL cold 70 % EtOH (made with DEPC-treated H_2O) and re-pelleted in a microcentrifuge (14,000 rpm, 18,000 × *g*) for 10 min at room temperature. The supernatant is aspirated and the tube placed on ice.
5. Glass micropipettes (Drummond) are fabricated using a gravity pipette puller (we use a Narishige PP-830). Micropipettes are fabricated on the day of use.
6. Each micropipette is viewed under a standard light microscope (10× objective) and the tip is broken using a sterilized glass stir rod, so that the tip diameter is approximately 20–30 μm (see Note 7).
7. We use a Nanoject II injector (Drummond) mounted on a manual micromanipulator (World Precision Instruments, M3301R).
8. The injector plunger is extended about 1/4 of its maximum distance. A glass micropipette is then filled with mineral oil (approximately 1 cm at the back end) using a syringe. The micropipette is then placed onto the injector plunger and seated into the retaining collar. Slowly, the plunger is extended until mineral oil is pushed to the tip of the glass pipette.
9. To serve as a reservoir for the RNA solution, a cap from a microcentrifuge tube is cut and placed underneath the dissection microscope.
10. The RNA pellet is resuspended in 1–2 μL DEPC-treated H_2O. The solution is incubated for 2 min in a 65 °C water bath and immediately pipetted into the microcentrifuge tube cap, in the corner furthest from the injector micropipette. Using the micromanipulator, the injector micropipette tip is carefully moved into the RNA droplet.
11. The micropipette is backfilled with RNA solution by slowly withdrawing the plunger (see Note 8).

12. 10–12 Oocytes are transferred to the injection stage with a sterile transfer pipette. Each oocyte will come to rest in a square of the mesh. The solution level should be adjusted so that the oocytes are only partially submerged. This keeps the oocytes from moving during injection.
13. While viewed through a dissection microscope, micropipette tip is moved until it slightly dimples the oocyte. The injector is tapped lightly to penetrate the oocyte. The oocyte is injected with 50 nL RNA solution. During a successful injection, the oocyte will swell slightly (see Note 9). The micropipette is withdrawn and the injection stage is moved to position the next oocyte. The injection procedure is repeated until all oocytes are injected.
14. Using a sterile transfer pipette, injected oocytes are transferred to a clean glass dish with Barth's regular solution and incubated at 18 °C.
15. After 1–2 h, any dead or unhealthy oocytes are removed and the Barth's regular solution is changed. Blemishes or discolorations indicate an unhealthy oocyte. Dead oocytes take on the appearance of marble.
16. Each day, the oocytes should be examined under a dissection microscope. Any dead or unhealthy oocytes should be removed.
17. Barth's regular solution is changed daily.
18. Oocytes can usually be maintained for 5–7 days.
19. Inclusion of an antibiotic in the Barth's regular solution is critical (see Note 10).

3.6 Electro physiological Assay of Odorant Receptors Expressed in Xenopus Oocytes

3.6.1 Odorant Preparation

1. Odorants are acquired from commercial suppliers at the highest purity available.
2. Odorants that are susceptible to oxidation should be stored under an inert gas, such as argon.
3. 0.5 or 1 M odorant stock solutions are prepared monthly in DMSO (occasionally in EtOH) and stored at -20 °C.
4. On the day of experimentation, working solutions are prepared in ND96.

3.6.2 Electrophysiology

1. Techniques for electrophysiological recording in *Xenopus* oocytes have been extensively described elsewhere (10).
2. A variety of integrated and automated systems are available for electrophysiological recording with oocytes. We use the OpusXpress 6000A (Molecular Devices), but this system is no longer commercially available. However, other systems are available, such as Robocyte and HiClamp (Multichannel Systems).

3.6.3 Mammalian ORs

1. An electrophysiological output from the activation of a mammalian OR is achieved by coupling this GPCR to the human cystic fibrosis transmembrane regulator (CFTR). Here is how it works. The $G\alpha_s$-like $G\alpha_{olf}$ is also expressed to allow OR activation to cause stimulation of endogenous adenylate cyclase. The resulting increase in cAMP stimulates endogenous protein kinase A, which phosphorylates the CFTR, a phosphorylation-dependent chloride channel (see Note 11). Our development of this method (4) is based on earlier work with adrenergic receptors (14).
2. Each oocyte is injected with 50 nL water containing 40 ng RNA encoding the mammalian OR, 10 ng RNA encoding human $G\alpha_{olf}$, and 2 ng RNA encoding the human CFTR.
3. Expression levels will be sufficient for experimentation approximately 2–3 days after injection (see Note 12).
4. Oocytes are clamped at −70 mV. Responses are sampled at 100 Hz and filtered at 20 Hz (4-pole, Bessel low pass, −3 dB).
5. Each odorant is applied for 15 s (at a flow rate of 1.65 mL/min) and the oocyte is washed for 10 min (at a flow rate of 3 mL/min) between odorant applications. 3-Isobutyl-1-methylxanthine (IBMX) can be applied (at 1 mM) to confirm function of the CFTR.
6. In Fig. 1a, we provide an example. MOR174-9 (mOR-EG) is a well-characterized mouse OR (15) that responds to eugenol and related compounds.
7. The electrophysiological response is slow, often initiating after the odorant has already been washed away and reaching a peak 30–90 s after the odorant application has ended (Fig. 1b). This is due to the time that is needed for the signal transduction pathway to culminate in the phosphorylation of the CFTR (see Note 13).

3.6.4 Insect ORs

1. Insect ORs are ligand-gated ion channels (16, 17). Thus, activation of these receptors in oocytes will directly result in an electrophysiological output.
2. A functional insect OR consists of two components, a variable odorant specificity subunit (of which there are many in each species) and a common subunit, Orco (see Note 14). The stoichiometry of these subunits in a functional insect OR is currently unknown.
3. Each oocyte is injected with 50 nL water containing 25 ng RNA encoding an odorant specificity subunit (such as Dmel\Or35a) and 25 ng RNA encoding the common subunit (such as Dmel\Orco).

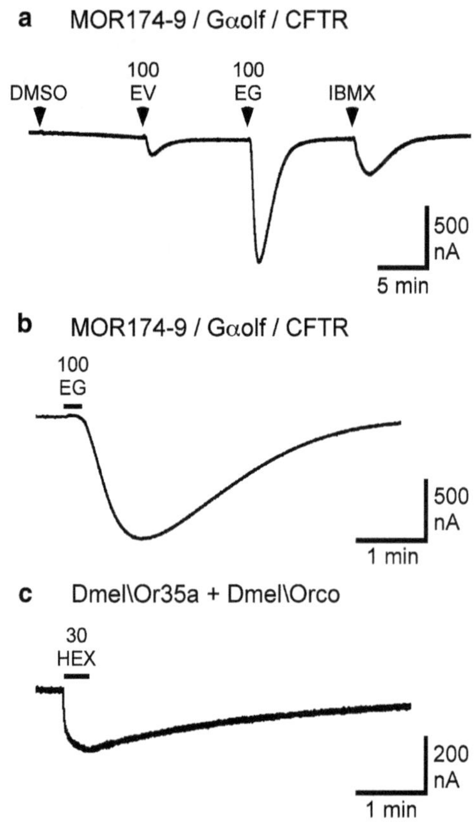

Fig. 1 Functional responses of mammalian and insect ORs expressed in *Xenopus* oocytes. (**a**) An oocyte expressing MOR174-9, $G\alpha_{olf}$, and CFTR is challenged with 15 s applications of 0.02 % DMSO (vehicle), 100 μM ethyl vanillin (EV), 100 μM eugenol (EG), and 1 mM IBMX. (**b**) An expansion of the time scale for the eugenol response of the oocyte shown in (**a**). (**c**) An oocyte expressing Dmel\Or35a and Dmel\Orco is challenged with a 20 s application of 30 μM hexanol

4. Expression levels will be sufficient for experimentation approximately 2–3 days after injection.
5. Oocytes are clamped at −70 mV. Responses are sampled at 100 Hz and filtered at 20 Hz (4-pole, Bessel low pass, −3 dB).
6. In Fig. 1c, we provide an example. Dmel\Or35a + Dmel\Orco is a well-characterized receptor from *Drosophila melanogaster* that responds to hexanol (18, 19).
7. Odorants can be applied for a variety of time periods. In our example, we are applying odorant for 20 s (at a flow rate of 1.65 mL/min). Oocytes should be washed extensively between odorant applications (10 min at a flow rate of 3 mL/min).
8. In contrast to the delayed response kinetics seen with mammalian ORs (Fig. 1b), an oocyte expressing an insect OR

responds immediately upon perfusion with odorant (Fig. 1c). The response begins to decline as soon as the odorant begins to be washed out.

4 Notes

1. Except for the "frog water" (which is made with RO water), all H_2O in our protocols is high-purity water from a Milli-Q system.
2. This is a crude preparation of collagenase, containing minor amounts of other proteases. The levels of these minor proteases, which are critical to the oocyte preparation technique, vary from batch to batch. Thus, each new batch of collagenase must be optimized, by trying several different concentrations.
3. We have found that while some mammalian ORs require the presence of a "rho-tag" for functional expression, others do not. It is unclear what accounts for this difference. As a general practice, we include the "rho-tag" for all mammalian OR expression studies, as the presence of this sequence does not appear to alter the function of mammalian ORs that do not require it.
4. When synthesizing, handling, and injecting RNA, appropriate precautions must be taken. All materials and reagents should be RNase-free. Aerosol-resistant filtered pipette tips should be used. All working surfaces, including the tube rack, should be thoroughly cleaned with 70 % EtOH.
5. Selected oocytes should have a perfectly spherical shape and a clear demarcation between the dark animal pole and light vegetal pole. Ideally, the demarcation between animal and vegetal poles should not contain an equatorial banded pattern, as this can indicate an overly mature oocyte. The severity of oocyte selection criteria is largely dependent on the overall quality of the batch. It should be noted that we find only a loose association between esthetically pleasing oocytes and expression capabilities. Therefore, we attempt to use almost every oocyte batch removed from a particular frog for experiments. The selected oocytes must be free of the follicle cell layer. Oocytes with an intact follicle cell layer will appear shiny and may have noticeable blood vessels. While oocytes with an intact follicle cell layer can be injected with RNA, recording pipettes will most likely break when electrophysiological recordings are attempted.
6. Gentle handling of RNA tubes is needed to avoid disturbance of the RNA pellet and accidental aspiration during washing steps. If using a pipette attached to a vacuum source for aspiration,

we recommend a borosilicate glass Pasteur pipette flamed and manually pulled to a fine tip. While some practice of this technique is usually required to achieve suitable pipette dimensions, we find this method can prevent aspiration of the pellet itself. Visualization of the RNA pellet after precipitation is not always possible and researchers should make note of the estimated location of the RNA pellet.

7. Ideally, the tips of RNA injection pipettes should be beveled to allow easy piercing of the oocyte vitelline membrane. This can be achieved by orienting the flat face of the end of the glass stir rod perpendicular to the tip of the injection pipette when fracturing the tip.

8. The resuspended RNA solution can be quite viscous and may clog the micropipette during loading. Quickly switching between filling and expelling the solution may improve loading. If air bubbles start to form within the micropipette, dispense solution and repeat procedure. Increasing the diameter of the micropipette tip can reduce the formation of air bubbles during filling.

9. We inject oocytes with RNA in either the animal or vegetal pole. The location of the injection site has no effect on receptor expression levels.

10. The primary infection problem for *Xenopus* oocytes is *Stenotrophomonas maltophilia*, an aerobic, gram-negative bacillus. *S. maltophilia* colonizes aquatic environments in hospitals and labs, making a *Xenopus* colony water system an ideal location. The bacterium appears to be transferred to oocyte cultures via *Xenopus* skin mucus. This is the reason for the stringent precautions during surgery (Subheading 3.2). If allowed to grow unchecked in the oocyte culture, *S. maltophilia* will kill the oocytes. The major problem here is that *S. maltophilia* can be resistant to many common antibiotics and antibiotic combinations, including Penicillin–Streptomycin, Gentamicin, Sulfamethoxazole–Trimethoprim, and Amikacin. We are currently using Ceftazidime, while keeping Ciprofloxacin in reserve.

11. *Xenopus* oocytes maintain a high intracellular chloride concentration and the reversal potential for chloride is approximately −35 mV. Thus, when the CFTR is activated at a holding potential of −70 mV, chloride ions will exit the cell, yielding an "inward" current (anion efflux mimics cation influx).

12. Expression levels can vary dramatically from oocyte to oocyte within one batch as well as from batch to batch (frog to frog as well as surgery to surgery with the same frog). In addition, there is a seasonal variation, with expression levels tending to be poorer in the summer. Proper attention to the water quality

of the frog aquarium system, the health of the frogs and the care of the oocytes, will help to minimize (but can't entirely eliminate) the frequency of "bad batches" of oocytes.

13. The CFTR can be directly activated by a wide variety of structures (20), making oocytes expressing the CFTR prone to false positives, particularly when using high concentrations of odorant (>100 μM). Thus, it is important to also test oocytes expressing the CFTR, but no OR, for potential false positives.
14. The naming of the common subunit in insect ORs has been variable across different insect species. For example, this subunit is Or83b in *Drosophila melanogaster*, Or2 in *Apis mellifera* (honey bee), and Or7 in various mosquito species. To resolve this confusion, a unified nomenclature has been proposed (21), referring to the common subunit in all species as the Orco (odorant receptor co-receptor). We are adopting this nomenclature here.

References

1. Mombaerts P (2004) Genes and ligands for odorant, vomeronasal and taste receptors. Nat Rev Neurosci 5:263–278
2. McClintock TS, Sammeta N (2003) Trafficking prerogatives of olfactory receptors. Neuroreport 14:1547–1552
3. Lu M, Staszewski L, Echeverri F, Xu H, Moyer BD (2004) Endoplasmic reticulum degradation impedes olfactory G-protein coupled receptor functional expression. BMC Cell Biol 5:34
4. Abaffy T, Matsunami H, Luetje CW (2006) Functional analysis of a mammalian odorant receptor subfamily. J Neurochem 97: 1506–1518
5. Hsiao B, Mihalak KB, Repicky SE, Everhart D, Mederos A, Malhotra A, Luetje CW (2006) Determinants of zinc potentiation on the α4 subunit of neuronal nicotinic receptors. Mol Pharmacol 69:27–36
6. Mihalak KB, Carroll FI, Luetje CW (2006) Varenicline is a partial agonist at α4β2 and a full agonist at α7 neuronal nicotinic receptors. Mol Pharmacol 70:801–805
7. Abaffy T, Malhotra A, Luetje CW (2007) The molecular basis for ligand specificity in a mouse olfactory receptor: a network of functionally important residues. J Biol Chem 282: 1216–1224
8. Hsiao B, Mihalak KB, Magleby KL, Luetje CW (2008) Zinc potentiates neuronal nicotinic receptors by increasing burst duration. J Neurophysiol 99:999–1007
9. Papke RL, Smith-Maxwell C (2009) High-throughput electrophysiology with *Xenopus* oocytes. Comb Chem High Throughput Screen 12:38–50
10. Sherman-Gold R (ed) (2008) The Axon guide a guide to electrophysiology & biophysics laboratory techniques, 3rd edn. MDS Analytical Technologies, Sunnyvale, CA
11. Liman ER, Tytgat J, Hess P (1992) Subunit stoichiometry of a mammalian K+channel determined by construction of multimeric cDNAs. Neuron 9:861–871
12. Krautwurst D, Yau KW, Reed RR (1998) Identification of ligands for olfactory receptors by functional expression of a receptor library. Cell 95:917–926
13. Saito H, Kubota M, Roberts RW, Chi Q, Matsunami H (2004) RTP family members induce functional expression of mammalian odorant receptors. Cell 119:679–691
14. Uezono Y, Bradley J, Min C, McCarty NA, Quick M, Riordan JR, Chavkin C, Zinn K, Lester HA, Davidson N (1993) Receptors that couple to 2 classes of G proteins increase cAMP and activate CFTR expressed in Xenopus oocytes. Receptors Channels 1: 233–241
15. Kajiya K, Inaki K, Tanaka M, Haga T, Kataoka H, Touhara K (2001) Molecular bases of odor discrimination: reconstitution of olfactory receptors that recognize overlapping sets of odorants. J Neurosci 21:6018–6025

16. Sato K, Pellegrino M, Nakagawa T, Nakagawa T, Vosshall LB, Touhara K (2008) Insect olfactory receptors are heteromeric ligand-gated ion channels. Nature 452:1002–1006
17. Wicher D, Schafer R, Bauernfeind R, Stensmyr MC, Heller R, Heinemann SH, Hansson BS (2008) Drosophila odorant receptors are both ligand-gated and cyclic-nucleotide-activated cation channels. Nature 452:1007–1011
18. Hallem EA, Ho MG, Carlson JR (2004) The molecular basis of odor coding in the Drosophila antenna. Cell 117:965–979
19. Wanner KW, Nichols AS, Walden KKO, Brockmann A, Luetje CW, Robertson HM (2007) A honeybee odorant receptor for the queen substance 9-oxo-2-decenoic acid. Proc Natl Acad Sci U S A 104:14383–14388
20. Ma T, Vetrivel L, Yang H, Pedemonte N, Zegarra-Moran O, Galietta LJ, Verkman AS (2002) High-affinity activators of cystic fibrosis transmembrane conductance regulator (CFTR) chloride conductance identified by high-throughput screening. J Biol Chem 277: 37235–37241
21. Vosshall LB, Hansson BS (2011) A unified nomenclature system for the insect olfactory co-receptor. Chem Senses 36(6):497–498, *Flybase*, FBrf0212857

Chapter 15

Pharmacology of Mammalian Olfactory Receptors

Richard S. Smith, Zita Peterlin, and Ricardo C. Araneda

Abstract

Mammalian species have evolved a large and diverse number of odorant receptors (ORs). These proteins comprise the largest family of G-protein-coupled receptors (GPCRs) known, amounting to ~1,000 different receptors in the rodent. From the perspective of olfactory coding, the availability of such a vast number of chemosensory receptors poses several fascinating questions; in addition, such a large repertoire provides an attractive biological model to study ligand–receptor interactions. The limited functional expression of these receptors in heterologous systems, however, has greatly hampered attempts to deorphanize them. We have employed a successful approach that combines electrophysiological and imaging techniques to analyze the response profiles of single sensory neurons. Our approach has enabled us to characterize the "odor space" of a population of native aldehyde receptors and the molecular range of a genetically engineered receptor, OR-I7.

Key words Olfactory receptors, Aldehydes, OR-I7, Sensory neurons, Ca-imaging

1 Introduction

In the vertebrate olfactory system, odorants are recognized by a combinatorial mechanism (1). A single receptor can recognize multiple odorants, and multiple receptors can detect the same odorant; indeed, odorants are recognized by different combinations of odorant receptors (ORs). The mammalian olfactory system employs a large number of receptors that comprise the largest family of G-protein-coupled receptors (GPCRs) known (2). The diversity of these receptors and the number of molecules they can potentially bind impose a great challenge to our understanding of peripheral olfactory processing (3). The difficulty of this challenge is exacerbated by our lack of knowledge of the mechanisms of OR expression, which limits the use of heterologous systems (4, 5). Thus, the initial excitement of the discovery of ORs in the early 1990 has slowly yielded to the realization that deorphanizing, the pairing of each of these receptors with a cognate ligand, will pose many challenges, key among which is

Chiquito J. Crasto (ed.), *Olfactory Receptors: Methods and Protocols*, Methods in Molecular Biology, vol. 1003,
DOI 10.1007/978-1-62703-377-0_15, © Springer Science+Business Media, LLC 2013

the paltry expression of these proteins in systems suitable for deorphanization. More than two-thirds of known ORs remain orphaned. To combat this, rather than relying on artificial expression systems we have used a different approach to determine the "odor space" of ORs, which is to analyze the response profiles of single sensory neurons (1, 6–8). The success of this approach relies on the peculiarity of OR expression—an olfactory sensory neuron expresses only one receptor from the available repertoire (2). Thus, we expose individual cells to a selected pool of odorants, with specific structural and chemical properties. The cell's responses are then analyzed using electrophysiological and imaging techniques. The odor response profile of a single cell is then, the odorant profile of a single OR. This approach allows the functional analyses of a large number of receptors in their native environment. By using a slightly different approach, sensory neurons have been previously used to express selected receptors using viral constructs containing an OR of interest (4, 9). The nasal epithelium is utilized as a proficient expression system and a single OR can be then challenged with a set of various odor molecules to determine its molecular range. We summarize below the nuances of the technique used to prepare isolated sensory neurons for calcium imaging recordings.

2 Materials

2.1 Isolation of Sensory Neurons

1. Ringer's solution: 138 mM NaCl, 5 mM KCl, 1 mM $CaCl_2$, 1.5 mM $MgCl_2$, 10 mM HEPES, 10 mM D-glucose, pH 7.4 (see Note 1).
2. Divalent-free Ringer's used for dissociation: 145 mM NaCl, 5.6 mM KCl, 10 mM HEPES, 10 mM D-glucose, 4 mM ethyleneglycol tetraacetic acid (EGTA), pH 7.4. This solution can be aliquoted and stored at 4 °C.
3. Enzyme mixture for sensory neurons dissociation: Divalent-free Ringer's containing 5 mg/mL bovine serum albumin, 1 mg/mL collagenase, 2.4 U/mL dispase II, and 100 μL deoxyribonuclease II (DNase II).
4. Concanavalin A (10 mg/mL in water)-coated glass coverslips (22 × 22 mm, see Note 2).
5. Culture medium; Dulbecco's Modified Eagle's Medium/F-12 (DMEM/F12) supplemented with 10 % fetal bovine serum (FBS), 1× insulin–transferrin–selenium-X, 2 mM glutamine, 100 U/mL penicillin, and 100 μg/mL streptomycin. This solution is filtered through a Nalgene filter flask and aliquoted into 50 mL portions.

2.2 Calcium Imaging Recordings

1. Loading solution: Fura-2AM plus pluronic acid (F-127, 80 μg/mL) in Ringer's (see Note 3).
2. Inverted fluorescence microscope (IMT-2 Olympus, Tokyo, Japan) equipped with a SIT camera (C2400-08, Hamamatsu Photonics, Hamamatsu, Japan) connected to a frame grabber (LG-3, Scion, Frederick, MD).
3. Modified recording chamber (Warner, Hamdem, CT) connected to a multichannel perfusion system (custom made, see Note 4).
4. The NIH Image software for data acquisition and analysis (NIH, Bethesda, MD) with customized macros written for the shutter control (Uniblitz, Vincent associates, Rochester, NY) and time-lapse imaging.
5. Odor solutions are prepared from pure odors in DMSO at a concentration of 1 mM and stored at 4 °C.

3 Methods

Olfactory sensory neurons are not well-suited for long-term cell cultures and quickly undergo apoptosis once their axon is severed. Therefore, great care should be exercised during the dissection and dissociation of the sensory epithelium. This protocol works well for both mice and rats.

3.1 Dissection of the Sensory Epithelium

1. Once the animal has been euthanized and the head isolated, the fur is removed and the head bisected from the midline, starting from the most rostral point, cutting caudally using the nasal bones as guideline.
2. The head halves are placed on a plastic 35 mm Petri dish and covered with ice-cold divalent-free Ringer. Use small curved scissors to remove the turbinates and septal bone from the skull and transfer them to a Petri dish.
3. The olfactory sensory tissue is then carefully dissected out from the turbinates and from the septal bone gently but quickly using a pair of fine-tip forceps. The whole procedure should be done in ice-cold Ringer's and over ice.

3.2 Dissociation of the Sensory Neurons

1. Place the tissue on a Petri dish containing divalent-free Ringer's and carefully mince using a microsurgery scissors.
2. Transfer the tissue to a Falcon plastic conical 15 mL tube and remove the excess of divalent-free Ringer's and add the enzyme mixture supplemented the DNase (100 μL).
3. Incubate the tissue in the enzyme mixture at 37 °C for 30 min (mouse) or 45 min (rat) on a shaker, at approximately 35 rpm.

4. After the incubation, place the tube inside a tissue culture hood and let a small piece of tissue settle at the bottom of the conical tube. Use a transfer pipette to suck off as much enzyme solution as possible and add 5 mL of pre-warmed culture media (37 °C). Gently mix with the tissue to stop the enzyme reaction and let the tissue settle at the bottom of the tube. Carefully transfer the tissue to another conical tube containing 5 mL of culture media, let the tissue settle and complete two additional washes.
5. Remove as much medium as possible and add 500 μL of fresh culture medium. The volume needed at this step can be varied depending on the amount of tissue; in general, rats require more volume.
6. For the dissociation, it works best to disperse the cells by holding the tube vertically and striking it downward on a firm surface. At this point, the solution becomes clouded as the cells have dissociated.

3.3 Fura-2AM Loading

1. Let the larger pieces of tissue settle and take 50 μL of the cell suspension from the supernatant; spot the cell suspension onto a Concanavalin A (Con A)-coated coverslip. Let the cells attach to the Con A for about 20 min.
2. Add 1.5 mL of media being careful not to disturb the spot of Con A containing the cells, and then using a suction pipette carefully remove the media from the dish, repeat this and wash about three times, avoiding to let the cells dry in between the washes.
3. After the last wash, add 1.5 mL of media containing ascorbic acid (100 μM) and incubate the cells for 45 min to 1 h.
4. Wash the dish twice with pre-warmed (37 °C) Ringer's and then add 1 mL of Ringer containing the Fura-2AM and incubate for 45 min at room temperature. To prepare the loading solution, thaw a tube containing a 6.25 μL aliquot and then add 1 mL Ringer and 1 μL of pluronic acid and vortex for at least 1 min.
5. After the 45 min incubation period remove, the Ringer's containing the Fura and wash twice with regular Ringer's. Leave in regular Ringer's for at least 10 min, before imaging the cells.

3.4 Calcium Imaging

1. The coverslip is mounted at the bottom of the perfusion chamber over a small spot of Vaseline used to prevent movements during the perfusion.
2. Recordings are made at 380 nm excitation and 510 nm emission (10). Images are acquired at a rate of 15 frames per minute. After a baseline of 40 s, odors are perfused into the

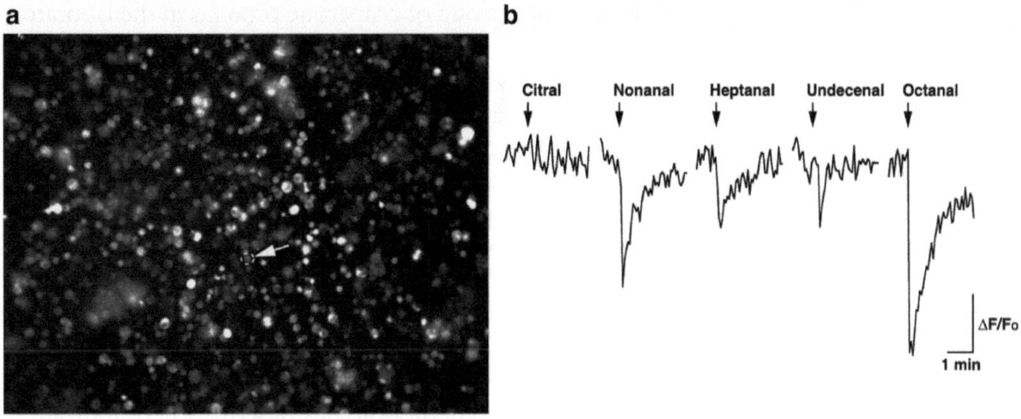

Fig. 1 Calcium imaging of olfactory sensory neurons. (**a**) Cells were loaded with Fura-2AM for 45 min. Data are analyzed offline by drawing a region of interest around responding cells (*white arrow*) and determining the change in fluorescent intensity. We find that, on average, 20 % of the cells in the recording field are viable olfactory sensory neurons, as measured by the response to a high KCl stimulus (50 mM) and IBMX (2 mM, not shown). (**b**) Responses to aliphatic aldehydes of different carbon lengths in the cell shown in (**a**). An odorant is considered to elicit a response when the change in $\Delta F/F_0$ is higher than two times the value of the standard deviation of the baseline and the decrease in $\Delta F/F_0$ lasts more than 20 s. The profile of this cell corresponds to that of OR-I7 (9)

chamber for 8 s in enough volume to completely replace the solution in the chamber (~200 μL). An example of Fura-2AM loaded cells is shown in Fig. 1a.

3. Odor responses are quantified as the fractional change in fluorescent light intensity: $\Delta F/F_0$ or $(F - F_0)/F_0$, where F is the fluorescent light intensity at each point and F_0 is the value of emitted fluorescent light before the odor application (baseline). An example of a cell's responses to aliphatic aldehydes of different carbon lengths is shown in Fig. 1b (see Note 5).
4. We use odorant at concentrations 0.3–300 μM, applied at intervals of at least 3 min. Odorant solutions are freshly prepared in Ringer's by dilutions from odorant stocks made in dimethyl sulfoxide (DMSO) and applied randomly. We routinely limit the number of odor presentations to less than 15 per cell.

3.5 Discussion

Since the discovery of ORs, more than 20 years ago (11), the deorphanization of ORs has been dishearteningly slow. A critical factor contributing to the slow progress of deorphanization is the poor expression of ORs in heterologous systems that are commonly used to study receptor pharmacology, such as transfected cell lines and expression in Xenopus oocytes. Early observations indicated that intracellular trafficking of ORs was impaired and several molecular approaches were undertaken to overcome this issue (12).

The identification of a group of chaperone proteins in the laboratory of Dr. Hiroaki Matsunami has offered a promising starting point (13). The researchers co-expressed these proteins with ORs in transfected cells and coupled their expression with a luciferase reporting system, which greatly contributed to the deorphanization of several ORs. This approach has proven limited however, as most ORs still have expression issues and a large number of ORs remains orphan. The prevailing view is that various cofactor proteins (or chaperones), yet to be discovered, are necessary for the appropriate intracellular trafficking and targeting of ORs to the cell's surface (4).

The advantage of recording responses directly from intact sensory neurons, as our methodologies described here indicate, is obvious; this is because the native transduction machinery is already present. In addition, compared with the kinetics of the responses in the luciferase assay (hours), the kinetics of the responses in the isolated cell preparation is closer to that of an in vivo environment (seconds). A small drawback of our approach, however, is that in these experiments isolated sensory neurons are surrounded by an artificial solution, instead of the natural mucus layer that naturally bathes their cilia. Therefore, the odorant–OR interaction is not under the ideal physiological conditions. Nevertheless, odor application is fast and several odors can be applied to the same cell allowing for the rapid screening of odorants and thorough characterization of an OR pharmacology (14). Lastly, a critical step is to follow the characterization of the OR pharmacological profile by its genetic identification, using single cell PCR of the recorded neuron. Only when the receptor sequence is known and reconstituted in vivo can the receptor's pharmacological profile be confirmed.

4 Notes

1. All solutions are prepared in ultrapure water (less than 18 MΩ cm resistivity). Unless indicated all solutions and materials are prepared under sterile conditions under a tissue culture hood.
2. The Concanavalin A solution is filtered and aliquoted into 100 μL stored at −20 °C.
3. The Fura-2AM vial is aliquoted into 6.25 μL in 1.5 mL Eppendorf tubes and stored at −20 °C.
4. The perfusion chamber is modified so that the perfusion solution only fills the diamond area. Odors are applied using syringes connected to the perfusion system via a manifold.

5. When studying the responses to different molecules we use one odor as a control to test the viability throughout the recordings, for example, octanal (6). Alternatively, responses can be normalized to the cell's response to the phosphodiesterase inhibitor 3-iso-butyl-1-xanthine (IBMX, 2 mM) or the activator of the adenylyl cyclase, Forskolin (10 μM) (14).

Acknowledgment

The authors would like to thank Xinmin Zhang for his helpful input in developing and optimizing these procedures.

References

1. Malnic B, Hirono J, Sato T, Buck LB (1999) Combinatorial receptor codes for odors. Cell 96:713–723
2. Mombaerts P (2004) Genes and ligands for odorant, vomeronasal and taste receptors. Nat Rev Neurosci 5:263–278
3. Reisert J, Restrepo D (2009) Molecular tuning of odorant receptors and its implication for odor signal processing. Chem Senses 34:535–545
4. Touhara K (2007) Deorphanizing vertebrate olfactory receptors: recent advances in odorant response assays. Neurochem Int 51: 132–139
5. Matsunami H, Mainland JD, Dey S (2009) Trafficking of mammalian chemosensory receptors by receptor-transporting proteins. Ann N Y Acad Sci 1170:153–156
6. Araneda RC, Peterlin Z, Zhang X, Chesler A, Firestein S (2004) A pharmacological profile of the aldehyde receptor repertoire in rat olfactory epithelium. J Physiol 555: 743–756
7. Kajiya K, Inaki K, Tanaka M, Haga T, Kataoka H, Touhara K (2001) Molecular bases of odor discrimination: reconstitution of olfactory receptors that recognize overlapping sets of odorants. J Neurosci 21:6018–6025
8. Kaluza JF, Breer H (2000) Responsiveness of olfactory neurons to distinct aliphatic aldehydes. J Exp Biol 203(Pt 5):927–933
9. Araneda RC, Kini AD, Firestein S (2000) The molecular receptive range of an odorant receptor. Nat Neurosci 3:1248–1255
10. Yuste R, Lanni F, Konnerth A (2000) Imaging neurons: a laboratory manual. Cold Spring Harbor Laboratory Press, New York
11. Buck L, Axel R (1991) A novel multigene family may encode odorant receptors: a molecular basis for odor recognition. Cell 65: 175–187
12. McClintock TS, Sammeta N (2003) Trafficking prerogatives of olfactory receptors. Neuroreport 14:1547–1552
13. Saito H, Kubota M, Roberts RW, Chi Q, Matsunami H (2004) RTP family members induce functional expression of mammalian odorant receptors. Cell 119:679–691
14. Peterlin Z, Li Y, Sun G, Shah R, Firestein S, Ryan K (2008) The importance of odorant conformation to the binding and activation of a representative olfactory receptor. Chem Biol 15:1317–1327

Chapter 16

Visualizing Olfactory Receptor Expression and Localization in *Drosophila*

Michael Saina and Richard Benton

Abstract

Odor detection and discrimination by olfactory systems in vertebrates and invertebrates depend both on the selective expression of individual olfactory receptor genes in subpopulations of olfactory sensory neurons, and on the targeting of the encoded proteins to the exposed, ciliated endings of sensory dendrites. Techniques to visualize the expression and localization of olfactory receptor gene products in vivo have been essential to reveal the molecular logic of peripheral odor coding and to permit investigation of the developmental and cellular neurobiology of this sensory system. Here, we describe methods for detection of olfactory receptor transcripts and proteins in the antennal olfactory organ of the fruit fly, *Drosophila melanogaster*, an important genetic model organism. We include protocols both for antennal cryosections and whole-mount antennae. These methods can be adapted for detection of receptor expression in other olfactory and gustatory tissues in *Drosophila*, as well as in the chemosensory systems of other insects.

Key words Olfaction, Olfactory receptor, Gene expression, *Drosophila melanogaster*, Insect, Antenna, Fluorescent in situ hybridization, Immunohistochemistry

1 Introduction

Olfactory receptors form the molecular interface between the external, odorous world and the internal, neural activity patterns that represent stimulus identity and intensity. These transmembrane receptors are encoded by large, highly divergent and evolutionarily dynamic gene families (1). In most animals, individual receptors are expressed in unique subpopulations of olfactory sensory neurons (OSNs) in the nose (2, 3). The encoded proteins are targeted to the ciliated endings of OSN dendrites; here, they are exposed to environmental odor molecules and define the physiological responsiveness of the host OSN to these stimuli (4–6).

Insects are experimentally accessible models to investigate olfaction, as their olfactory circuits display the same organizational logic as those of vertebrates, but with significantly fewer receptors

Chiquito J. Crasto (ed.), *Olfactory Receptors: Methods and Protocols*, Methods in Molecular Biology, vol. 1003,
DOI 10.1007/978-1-62703-377-0_16, © Springer Science+Business Media, LLC 2013

and neurons (7). The fruit fly, *Drosophila melanogaster*, has emerged as a particularly powerful model, offering the possibility to combine molecular genetic approaches with neurophysiological and behavioral analyses (8, 9). The major olfactory organ in *Drosophila* is the third segment of the antenna (hereafter referred to as the "antenna"), a head appendage that is covered with porous sensory hairs, called sensilla. Sensilla fall into three main morphological classes (basiconic, trichoid, and coeloconic), and house the ciliated, dendritic endings of 1–4 OSNs (10).

Two repertoires of olfactory receptor genes are expressed in the *Drosophila* antenna: Odorant Receptors (ORs) (4), which encode 7-pass transmembrane proteins that are likely to function as odor-gated ion channels (11), and the Ionotropic Receptors (IRs) (12), which are related to ionotropic glutamate receptors (13). In the antenna, most individual ORs (~40) are expressed together with a common co-receptor subunit (ORCO) (14, 15), in unique subpopulations of OSNs innervating basiconic and trichoid sensilla (16, 17). Similarly, most IRs (~14) are expressed in specific coeloconic OSN subpopulations (12), together with one or more of the broadly expressed co-receptors (IR8a, IR25a, and IR76b) (18). Both receptor classes are selectively targeted to sensory cilia, where they mediate odor-evoked neuronal activity (19–21).

The relatively small number of olfactory receptors in *Drosophila* has permitted comprehensive description of their expression patterns and ligand specificity (16, 17, 22, 23), providing the most complete knowledge of peripheral odor coding mechanisms in any animal. Nevertheless, many outstanding issues remain, including the regulatory mechanisms underlying the selective OSN expression of individual receptors (2), the cellular processes and dynamics by which these proteins are targeted to (and removed from) cilia, the biochemistry of odor ligand–receptor interactions and how these lead to neuronal depolarization. Moreover, comparative studies of olfactory receptor expression and function in *Drosophila melanogaster*, other drosophilids and more distantly related insects have stimulated significant interest in the evolution of the peripheral olfactory system (24, 25). As such, analysis of olfactory receptor expression and localization remain central techniques in the field.

In this chapter, we describe methods for analysis of olfactory receptor gene expression by RNA fluorescent in situ hybridization (FISH) and receptor protein localization by immunohistochemistry (IHC) in the *Drosophila* antenna. We present in detail established protocols for analysis of antennal cryosections (14, 15, 17), as well as refined methods that permit analysis of whole-mount antennae. We provide accompanying images for the specialized apparatus for tissue collection in Fig. 1, and sample results for each

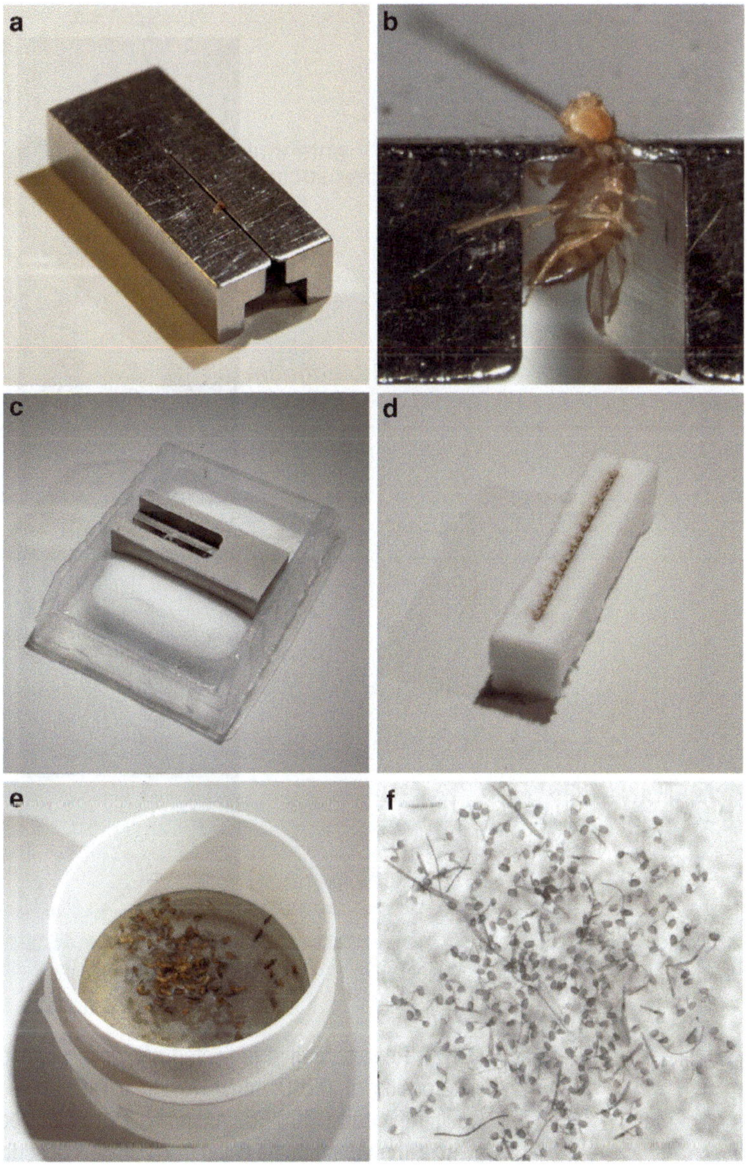

Fig. 1 Collection of *Drosophila* antennal sections and whole-mount antennae. (**a**) Fruit fly collar. (**b**) Side view of a collar illustrating the insertion orientation of a single fly; up to 15–20 animals can be accommodated within a single collar. (**c**) Collar inverted in the base mold, filled with OCT (after freezing). (**d**) Trimmed block of OCT containing fly heads. (**e**) Minisieve containing flies. (**f**) Antennal-enriched appendages collected through the 80 μM minisieve

of the histological techniques introduced in Fig. 2. These protocols can be easily adapted for detection of receptor expression in other types of chemosensory organ, such as the maxillary palp (a second olfactory organ) or gustatory neurons in the labellum of the proboscis (10), as well as in other insect species.

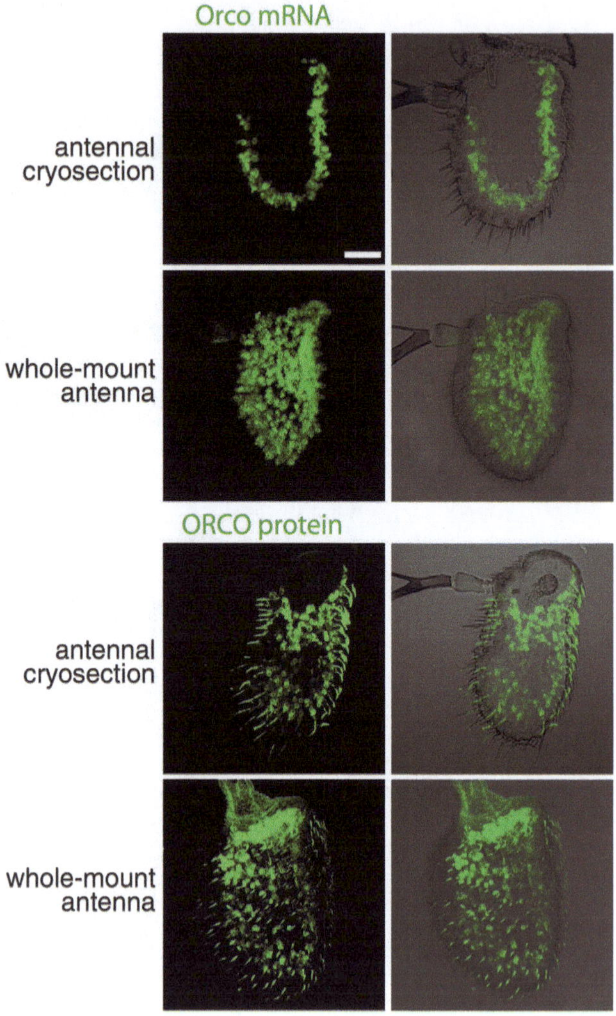

Fig. 2 Representative projected confocal microscope image stacks illustrating expression of ORCO mRNA (using a DIG-labeled riboprobe (14)) and protein (using the anti-ORCO EC2 antibody (15)) visualized in antennal cryosections and whole-mount antennae using the protocols described in this chapter. Fluorescent signals alone are shown on the *left* and overlaid on a brightfield image on the *right*. The scale bar represents 20 μm

2 Materials

2.1 Synthesis of Labeled RNA Probes

1. Plasmid vector (e.g., pGEM®-T or pBluescript) containing part or the entire coding region of the receptor gene of interest (typically 500–2,000 bp), flanked by bacteriophage RNA polymerase transcription initiation sites (i.e., T7, T3, or SP6).
2. Restriction enzyme and buffer (for linearization of the plasmid vector).

3. T7, T3, or SP6 polymerase transcription systems (contains buffer and Dithiothreitol (DTT)) (Promega Corp., Fitchburg, WI, USA; http://www.promega.com).
4. Digoxigenin (Dig) RNA labeling mix or Fluorescein RNA labeling mix (see Note 1).
5. RNasin®.
6. QIAquick PCR Purification Kit (see Note 2).
7. Illustra ProbeQuant™ columns.

2.2 Harvesting, Cryosectioning, and Fixation of Antennae for FISH and IHC

1. Custom fruit fly collar (#10731, 4M Instrument and Tool LLC, Islandia, NY, USA; http://www.4mprototype.com) (Fig. 1a).
2. Disposable base mold (#M475-4, Simport Scientific, Canada; http://www.simport.com).
3. Liquid nitrogen and an aluminum block that fits into the liquid nitrogen container; alternatively, dry ice, ethanol and a small plastic container containing a low grill rack.
4. O.C.T.™ Compound (Sakura Finetek Europe B.V., Alphen aan den Rijn, The Netherlands; http://www.sakuraeu.com).
5. Superfrost microscope slides.
6. Phosphate buffered saline (PBS) (10×): 1.54 M NaCl, 120 mM Na_2HPO_4, 90 mM NaH_2PO_4 (adjust to pH 7.4 with HCl, if necessary) and autoclave before storage at room temperature. Prepare a working solution by dilution of one part 10× stock with nine parts water (see Note 3).
7. Fixation solution: 4 % paraformaldehyde in PBS, pH 7.4. This should be prepared freshly, or frozen aliquots thawed immediately before use. Paraformaldehyde is toxic and potentially carcinogenic; it should only be handled with gloves and disposed of in an appropriate waste container.

2.3 RNA FISH on Antennal Cryosections

1. Phosphate buffered saline (PBS) (10×): 1.54 M NaCl, 120 mM Na_2HPO_4, 90 mM NaH_2PO_4 (adjust to pH 7.4 with HCl, if necessary) and autoclave before storage at room temperature. Prepare a working solution by dilution of one part 10× stock with nine parts water.
2. Acetylation solution: 1.85 g triethanolamine hydrochloride, 224 μL 10 N NaOH in 100 mL water. Immediately before use, add 0.25 mL acetic anhydride and shake the solution vigorously. Triethanolamine-HCl and acetic anhydride are both irritants and should only be handled with gloves and disposed of in appropriate waste containers.
3. Hybridization buffer: 50 % deionized formamide, 5× SSC, 0.5 mg/mL herring sperm DNA, 250 μg/mL yeast tRNA, 5× Denhart's Reagent, 0.1 % Tween-20, 2.5 mM EDTA, 0.25 %

CHAPS, 50 μg/mL heparin. It is better to make freshly, but can be stored at −20 °C. Formamide is teratogenic, highly corrosive to skin and eyes, and can cause death when ingested. It should only be handled with gloves and disposed of in an appropriate waste container.

4. Slide mailers.
5. Coverslips (25 × 60 mm).
6. 20× SSC stock: 3 M NaCl, 300 mM sodium citrate, pH 7.0.
7. 5× SSC.
8. 0.2× SSC.
9. TN buffer: 100 mM Tris–HCl, 150 mM NaCl, pH 7.0.
10. TNT buffer: 100 mM Tris–HCl, 150 mM NaCl, 0.05 % Tween-20, pH 7.0.
11. TNB: TNT buffer + 0.5 % Blocking Reagent (Roche, Basel, Switzerland). Make freshly or keep frozen until use.
12. Anti-Dig-Peroxidase (POD) or anti-Fluorescein-POD (Roche, Basel, Switzerland; http://www.roche.com).
13. Only in the case of combined FISH/IHC: primary and secondary antibodies of choice. We use secondary antibodies coupled to Alexa488, Cy3, or Cy5 fluorophores (Jackson Immuno Research, USA; http://www.jacksonimmuno.com).
14. Tyramide Signal Amplification System: TSA™ Plus-Fluorescein System or TSA™ Plus-Cy5 System (Perkin Elmer, Inc., Waltham, MA, USA; http://www.perkinelmer.com).
15. Vectashield mounting medium (Vector Laboratories, Inc. Burligame, CA, USA; http://www.vectorlabs.com).
16. Nail polish.

2.4 IHC on Antennal Cryosections

1. Slide mailers.
2. Coverslips (25 × 60 mm).
3. Phosphate buffered saline (PBS) (10×): 1.54 M NaCl, 120 mM Na_2HPO_4, 90 mM NaH_2PO_4 (adjust to pH 7.4 with HCl, if necessary) and autoclave before storage at room temperature. Prepare a working solution by dilution of one part 10× stock with nine parts water.
4. PBT: PBS + 0.1 % Triton X-100.
5. Blocking solution: 5 % heat-inactivated Normal Goat Serum in PBT. Inactivate the goat serum by heating it to 56 °C for 1 h before dilution.

Keep samples frozen until use.

6. Primary and secondary antibodies. We use secondary antibodies coupled to Alexa488, Cy3, or Cy5 fluorophores (Jackson Immuno Research, USA; http://www.jacksonimmuno.com).

7. Vectashield mounting medium (Vector Laboratories, Inc. Burligame, CA, USA; http://www.vectorlabs.com).
8. Nail polish.

2.5 Harvesting and Fixation of Whole-Mount Antennae for FISH and IHC

1. Petri dish.
2. Minisieve, mesh-width 80 μM (#378460000, bel-art products, NJ, USA; http://www.belart.com) (Fig. 1e).
3. Liquid nitrogen.
4. Phosphate buffered saline (PBS) (10×): 1.54 M NaCl, 120 mM Na_2HPO_4, 90 mM NaH_2PO_4 (adjust to pH 7.4 with HCl, if necessary) and autoclave before storage at room temperature. Prepare a working solution by dilution of one part 10× stock with nine parts water.
5. Fixation solution: 4 % paraformaldehyde in PBS, pH 7.4 + 3 % Triton X-100. This should be prepared freshly, or frozen aliquots thawed immediately before use. Paraformaldehyde is toxic and potentially carcinogenic; it should only be handled with gloves and disposed of in an appropriate waste container.

2.6 RNA FISH on Whole-Mount Antennae

1. Phosphate buffered saline (PBS) (10×): 1.54 M NaCl, 120 mM Na_2HPO_4, 90 mM NaH_2PO_4 (adjust to pH 7.4 with HCl, if necessary) and autoclave before storage at room temperature. Prepare a working solution by dilution of one part 10× stock with nine parts water.
2. PBS + 3 % Triton X-100.
3. PBT: PBS + 0.1 % Triton X-100.
4. Triethanolamine stock solution (10×): 1 M TEA in PBS, pH 7.4. TEA is a strong base and can be irritating and cause inflammatory reactions. It is suspected to be carcinogenic and toxic to aquatic environments; it should only be handled with gloves and disposed of in an appropriate waste container. Prepare a working solution (0.1 M) by diluting one part stock solution with nine parts PBT.
5. Acetic anhydride. This chemical is an irritant and flammable; it should only be handled with gloves and disposed of in an appropriate waste container.
6. 20× SSC stock: 3 M NaCl, 300 mM sodium citrate, pH 7.0.
7. 5× SSCT: 5× SSC + 0.1 % Triton X-100.
8. 2× SSCT: 2× SSC + 0.1 % Triton X-100.
9. Hybridization buffer: 50 % deionized formamide, 5× SSC, 0.5 mg/mL herring sperm DNA, 1× Denhart's reagent, 0.1 % Triton X-100, 3 mM EDTA. It is better to make freshly, but can be stored at −20 °C. Formamide is teratogenic, highly

corrosive to skin and eyes, and can cause death when ingested. It should only be handled with gloves and disposed of in an appropriate waste container.

10. TNTX buffer: 100 mM Tris–HCl, 150 mM NaCl, 0.1 % Triton X-100, pH 7.0.
11. TNB buffer: TNTX buffer + 0.5 % Blocking Reagent (Roche, Basel, Switzerland). This should be prepared freshly, or frozen aliquots thawed immediately before use.
12. Anti-Dig-POD or anti-Fluorescein-POD (Roche, Basel, Switzerland; http://www.roche.com).
13. For combined RNA FISH and IHC: primary and secondary antibodies of choice. We use secondary antibodies coupled to Alexa488, Cy3, or Cy5 fluorophores (Jackson Immuno Research, USA; http://www.jacksonimmuno.com).
14. Tyramide Signal Amplification System: TSA™ Plus-Fluorescein System or TSA™ Plus-Cy5 System (Perkin Elmer, Inc., Waltham, MA, USA).
15. Methanol.
16. BB:BA: Benzylbenzoate:Benzylalcohol, mixed 2:1 (v/v). These chemicals can be irritating and toxic; they should only be handled with gloves and disposed of in an appropriate waste container. Alternatively, Vectashield mounting medium (Vector Laboratories, Inc. Burlingame, CA, USA; http://www.vector-labs.com) could be used.
17. Microscope slides.
18. Coverslips (18 × 18 mm).
19. Nail polish.

2.7 IHC on Whole-Mount Antennae

1. Phosphate buffered saline (PBS) (10×): 1.54 M NaCl, 120 mM Na_2HPO_4, 90 mM NaH_2PO_4 (adjust to pH 7.4 with HCl, if necessary) and autoclave before storage at room temperature. Prepare a working solution by dilution of one part 10× stock with nine parts water.
2. PBS + 3 % Triton X-100.
3. PBT: PBS + 0.1 % Triton X-100.
4. Blocking solution: 5 % heat-inactivated Normal Goat Serum in PBT. Inactivate the goat serum by heating it to 56 °C for 1 h before dilution. Keep samples frozen until use.
5. Methanol.
6. BB:BA: Benzylbenzoate:Benzylalcohol, mixed 2:1 (v/v). These chemicals can be irritating and toxic; they should only be handled with gloves and disposed of in an appropriate waste container. Alternatively, Vectashield mounting medium (Vector

Laboratories, Inc. Burlingame, CA, USA; http://www.vector-labs.com) could be used.

7. Primary and secondary antibodies. We use secondary antibodies coupled to Alexa488, Cy3, or Cy5 fluorophores (Jackson Immuno Research, USA; http://www.jacksonimmuno.com).
8. Microscope slides.
9. Coverslips (18 × 18 mm).
10. Nail polish.

3 Methods

3.1 Template Preparation and Synthesis of Labeled RNA Probes

To detect endogenous transcripts, labeled, antisense RNA probes are required. These are prepared by linearizing a plasmid vector containing the coding sequence of the receptor of interest at a position in the vector that permits subsequent in vitro transcription of antisense RNA in the presence of a labeled nucleotide (e.g., Dig-UTP or Fluorescein-UTP) (see Note 1). DNA template purity and avoidance of contamination with RNases (e.g., using RNase-free plasticware and gloves) is critical for successful probe synthesis and stability.

1. Determine the sequence and orientation of the insert in the vector by sequencing, and chose the appropriate polymerase to generate your antisense probe.
2. Linearize the plasmid vector: digest the equivalent of 1 μg of insert DNA in 70 μL with 20 U of enzyme for 3 h.
3. Check 1 μL of the digest on an agarose gel to verify that the DNA is fully linearized.
4. Purify the digestion reaction using QIAquick PCR Purification Kit (see Note 2) according to the manufacturer's instructions. Elute in 35 μL water.
5. Check 1 μL of DNA on an agarose gel to ensure that the template is still present (see Note 4).
6. To 10.5 μL of template in water add, in order, for a total volume of 20 μL:

   4 μL Promega 5× Transcription Buffer, 2 μL Dig (or Fluorescein) RNA labeling mix, 1 μL 0.1 M DTT, 0.5 μL Promega RNAsin, 2 μL RNA polymerase. (The remaining DNA template can be stored at −20 °C for future transcription reactions).
7. Mix gently, spin down at low speed, and incubate at 37 °C for 2 h.
8. Purify probes over Illustra ProbeQuant™ columns according to the manufacturer's instructions (see Note 2).

9. Check 1 μL of probe on an agarose gel to ensure that the transcription reaction is successful (see Note 5).
10. Determine the probe concentration by spectrophotometry.
11. Add hybridization buffer (see Subheading 2.2) to a volume of 100 μL and store at −20 °C. In our experience, probes are stable in these conditions for up to several years.

3.2 Harvesting, Cryosectioning and Fixation of Antennae for FISH and IHC

1. Anesthetize the flies with CO_2 or on ice, and use forceps and a paintbrush to slide them into the fruit fly collar, with only the heads projecting (Fig. 1a, b).
2. Drizzle a line of OCT over the heads, and fill the base mold with OCT. Invert the collar in the mold so that the fly heads are completely submerged (Fig. 1c).
3. Chill rapidly by placing the base mold on an aluminum block precooled in liquid nitrogen, or in an ethanol/dry ice bath. In the latter case, be very careful to avoid contact of OCT with ethanol as it will dissolve; we place the mold on a metal grill standing in the bath so that only its bottom surface contacts the ethanol. Once the OCT compound is frozen, it will appear opaque white (Fig. 1c); this will take a few minutes.
4. Prize the collar off carefully, mold and, using a razor blade, trim the block of OCT (containing now only the fly heads) to leave maximum 1–2 mm border surrounding the line of heads (Fig. 1d). Take care to avoid the blade slipping by allowing the OCT to soften very slightly before cutting. If it becomes too soft, place it (handling with chilled tweezers) on dry ice briefly. The block of fly heads in OCT can be stored in Eppendorf tubes at −80 °C for at least several weeks.
5. Mount the sample on the holder of your cryostat and collect 14 μM sections on a Superfrost slide, according to the particular specifications of your cryostat. For antennae, you can discard most of the initial sections of the head, which will contain only the posteriormost regions.
6. For FISH, allow the sections to dry for 30 min and then submerge in fixation solution (e.g., in a 50 mL conical tube) for 10 min at room temperature (all steps in all protocols are at room temperature unless otherwise noted). For IHC, put the slide into fixation solution for 10 min immediately after sectioning.

3.3 RNA FISH on Antennal Cryosections

The RNA FISH method described is based upon the Tyramide Signal Amplification (TSA™) method (Perkin Elmer). In brief, (1) the labeled riboprobes are hybridized to endogenous transcripts, (2) a peroxidase-coupled antibody that recognizes the probe label is applied, (3) a non-fluorochromic substrate is added, which is converted to a fluorescent deposit by the RNA-bound peroxidase. This two-step amplification mechanism allows highly sensitive detection of transcripts in situ. Nevertheless, success in RNA FISH is dependent upon many factors, including the expression level of the gene to be detected, probe concentration, probe length and sequence, tissue fixation, hybridization temperature and the general quality of the reagents used. Such parameters may have to be optimized to detect your gene of interest. RNA molecules are susceptible to biochemical destruction by RNases, which are abundant enzymes in the environment, or to chemical destruction by metal ions or high pH; caution and precision are therefore required throughout the protocol.

1. Remove the slide from fixation solution and wash 3 × 3 min in PBS.
2. Incubate for 10 min in acetylation solution.
3. Wash 3 × 5 min in PBS.
4. Prehybridize by placing slides vertically in slide mailers filled with hybridization buffer at 65 °C for at least 1 h.
5. Prepare in a tube 100 μL hybridization buffer per slide and add the amount of probe you intend to use (see Note 6). Combinations of differentially labeled probes can also be used (see Note 13).
6. Denature the probe by heating to 80 °C for 5 min in a hot-block, and then chill on ice briefly.
7. Remove the slide from the pre-hybridization buffer; allow excess buffer to drain (see Note 7). Pipette the probe carefully onto the slide and add a coverslip to ensure an even distribute of the probe across the section and avoid dehydration by evaporation. Place horizontally in slide mailer, using separate mailer for slides with different probes to avoid cross-contamination. Place the mailer in a humidified box (see Note 8), and the box in a 65 °C oven. Hybridize for 16–24 h.
8. Remove the slides from the mailer and let the coverslips slide off by soaking the slides in 5× SSC prewarmed to 65 °C.
9. Wash 3 × 20 min in 0.2× SSC at 65 °C.
10. Equilibrate in TN for 5 min (at room temperature).
11. Block in TNB for 30 min in a vertical slide mailer.
12. Dilute anti-Dig-POD (or anti-Fluorescein-POD for Fluorescein-labeled probes) 1:500 in TNB; prepare 100 μL of antibody solution per slide. Pipette carefully on to the sections, add a coverslip, and place slides in a humidified chamber for 30 min (see Notes 7 and 8). It is possible to combine FISH with IHC at this step; simply add the desired primary antibody together with the anti-Dig/Fluorescein-POD, and incubate overnight at 4 °C.

13. Discard the coverslip (it should simply slide off if you hold the slide vertically) and wash slides 3×5 min in TNT.
14. Prepare the TSA™ Plus-Fluorescein (or TSA™ Plus-Cy5) solution freshly according to the manufacturer's guidelines and apply 150 μL per slide. Incubate for 10 min in the dark.
15. Wash 3×5 min in TNT in the dark.
16. If you are combining FISH with IHC: apply the appropriate secondary antibody in TNB, incubate in the dark for 2 h, and wash 3×5 min in TNT in the dark.
17. Add 60 μL Vectashield to each slide, overlay a coverslip, and, if desired, seal with nail polish. Slides should be stored at 4 °C in the dark.

3.4 IHC on Antennal Cryosections

1. After fixation, wash 3×15 min in PBS in a vertical slide mailer.
2. Incubate in blocking solution for 30 min.
3. Dilute the primary antibody as appropriate in blocking solution. Drain excess liquid from the slide (see Note 7), and carefully pipette on the antibody solution (100–500 μL, depending upon how much primary antibody stock you have; for small volumes (<200 μL), add a coverslip to ensure even distribution of antibody across the sections and to avoid drying). Place the slide in a humidified box at 4 °C overnight (see Note 8).
4. Discard the coverslip (it should simply slide off if you hold the slide vertically) and wash 3×15 min in PBT in a vertical slide mailer.
5. Dilute the secondary antibody as appropriate in blocking solution. Drain excess liquid from the slide (see Note 7), and carefully pipette on the antibody solution (100–500 μL; add a coverslip for volumes <200 μL). Place the slide in a humidified box for 2 h (see Note 8).
6. Wash 3×20 min in PBT.
7. Drain excess liquid off the slide. Add 60 μL Vectashield, overlay a coverslip, and, if desired, seal with nail polish. Slides should be stored at 4 °C in the dark.

3.5 Harvesting and Fixation of Whole-Mount Antennae for FISH and IHC

1. Put 4 mL of fixation solution in the petri dish; keep a 2 mL tube with 1.8 mL fixation solution on ice.
2. Anesthetize the flies with CO_2, and transfer them to the minisieve, precooled in liquid nitrogen (Fig. 1e).
3. Hold the minisieve in liquid nitrogen for 10 s to snap-freeze the flies.
4. Gently tap and shake the minisieve above the petri dish containing the fixation solution. The antennal segments will break

off and fall through the sieve into the fixative (Fig. 1f). Some other body parts might fall through the sieve, but these can be removed from the sample if necessary.

5. Under a binocular microscope, use a 200 μL pipette to collect and transfer the antennal segments into the 2 mL tube with fixation solution. If only a few flies are available, antennae can alternatively be picked off flies manually with forceps and directly placed in this fixation solution.
6. Fix for 2 h at 4 °C on a rotator set at low speed (<20 rpm). If desired, fixed antennae can be stored at this point for future staining (see Note 9).

3.6 RNA FISH on Whole-Mount Antennae

See Subheading 3.3 for a brief theoretical introduction. All washing/equilibration steps are performed with 500 μL solution unless otherwise noted. Do not invert the tubes during the washing steps as antennae may stick to the wall or lid and become dehydrated. Simply remove the solution in the tube carefully with a pipette (see Note 10), add the new solution, and allow the tube to stand to permit the antennae to settle to the bottom again before the next wash.

1. Wash 2 × 5 min in 2 mL PBS + 3 % Triton X-100.
2. Wash 3 × 5 min in PBT.
3. Rinse briefly in 0.1 M triethanolamine solution.
4. Acetylate for 10 min with triethanolamine solution containing 5 μL acetic anhydride/mL (add the acetic anhydride immediately before use and make sure it is dissolved).
5. Wash 2 × 5 min with PBT.
6. Exchange PBT with hybridization buffer and equilibrate for 10 min.
7. Exchange hybridization buffer with new one and prehybridize at 60 °C for at least 1 h.
8. Prepare probe solution by adding the Digoxigenin-labeled (or Fluorescein-labeled) riboprobe to 60 μL of prewarmed hybridization buffer to the desired final concentration (see Note 6). Combinations of differentially labeled probes can also be used (see Note 13). Hybridize for 36–72 h at 60 °C.
9. Exchange the probe solution with 5× SSCT and incubate for 10 min at 60 °C.
10. Wash with 5× SSCT for 10 min at 60 °C.
11. Wash 3 × 20 min with 2× SSCT at 60 °C.
12. Wash with TNTX buffer for 5 min (at room temperature).
13. Block for 2 h in 100 μL TNB.

14. Prepare 50 μL of antibody solution/sample by diluting anti-Dig-POD (or anti-Fluorescein-POD for Fluorescein-labeled probes) 1:350 in TNB and replace the blocking solution with the antibody solution. It is possible to combine FISH with IHC at this step; simply add the desired primary antibody together with the anti-Dig/Fluorescein-POD. Incubate overnight at 4 °C.
15. Wash 5 × 15 min in TNTX.
16. Prepare the TSA™ Plus-Fluorescein (or TSA™ Plus-Cy5) solution freshly according to the manufacturer's guidelines and apply 40 μL per tube. Incubate for 60–80 min in the dark.
17. Wash 5 × 10 min in TNTX buffer.
18. If you are combining FISH with IHC: incubate with secondary antibody diluted in 5 % normal goat serum in TNTX for 2–3 h in the dark. Wash 4 × 20 min in TNTX in the dark.
19. Dehydrate the antennae in methanol 3 × 5 min (see Note 11).
20. Remove as much methanol as possible without letting the antennae become dry. Add 40 μL BB:BA and mix well by pipetting.
21. Use nail polish to paint a 12 × 12 mm (inside dimension) frame on the slide; this provides just enough thickness to avoid the antennae being crushed by the coverslip. Wait until the nail polish has dried completely, mount the samples within this frame and add a coverslip (see Note 12). Do not use nail polish to seal the coverslip on the slide, as this will form an emulsion with the BB:BA. Slides should be stored at 4 °C in the dark.

3.7 IHC on Whole-Mount Antennae

All washing/equilibration steps are performed with 500 μL solution unless otherwise noted. Do not invert the tubes during the washing steps as antennae may stick to the wall or lid and become dehydrated. Simply remove the solution in the tube carefully with a pipette (see Note 10), add the new solution, and allow the tube to stand to permit the antennae to settle to the bottom again before the next wash.

1. After fixation, wash 2 × 15 min in 2 mL PBS + 3 % Triton X-100.
2. Wash 3 × 20 min in PBT.
3. Incubate in 100 μL blocking solution for 1–2 h.
4. Replace the blocking solution with primary antibody of choice diluted in blocking solution and incubate overnight at 4 °C. We recommend using the antibodies at slightly higher concentrations than on cryosections, to facilitate their deeper permeation of the whole-mount tissue.
5. Wash 5 × 20 min in PBT.

6. Incubate with secondary antibodies diluted in blocking solution for 2–3 h in the dark.
7. Wash 5 × 20 min in PBT.
8. Dehydrate the antennae in methanol 3 × 5 min (see Note 11).
9. Remove as much methanol as possible without letting the antennae become dry. Add 40 μL BB:BA and mix well by pipetting.
10. Use nail polish to paint a 12 × 12 mm (inside dimension) frame on the slide; this provides just enough thickness to avoid the antennae being crushed by the coverslip. Wait until the nail polish has dried completely, mount the samples within this frame and add a coverslip (see Note 12). Do not use nail polish to seal the coverslip on the slide, this will form an emulsion with the BB:BA. Slides should be stored at 4 °C in the dark.

4 Notes

1. Although Fluorescein has endogenous fluorescence, we find it is too weak after the FISH protocol for direct detection. We therefore simply use Fluorescein as an epitope tag, like Digoxigenin, which can be detected with anti-Fluorescein-Peroxidase coupled antibodies.
2. Although we note specific commercial kits here, for standard molecular biological manipulations, several alternatives are possible.
3. Unless stated otherwise, all solutions should be prepared in water that has a resistivity of 18.2 MW-cm and total organic content of less than five parts per billion. This standard is referred to as "water" in this text.
4. You can also prepare templates for probe synthesis by PCR, using one primer that is upstream of the polymerase binding site in the plasmid vector and another at the end of your insert. Purify the PCR product by standard phenol–chloroform extraction or with the QIAquick PCR Purification Kit. Determine the concentration and purity of the product with a spectrophotometer and proceed with the transcription reaction as for linearized plasmid templates using 0.5–1 mg/kb of purified PCR product, adding water to complete the reaction volume if necessary.
5. The in vitro transcribed RNA should appear as a new band on an agarose gel, distinct from the DNA template band. The size is not easily predicted using a DNA ladder because secondary structure in RNA can significantly affect its migration.
6. The concentration of the riboprobe required for detection of endogenous transcript must usually be experimentally

determined as it is dependent on the sequence and quality of the probe as well as on the expression level of the receptor gene. Typically, we use probe concentrations of 0.01–10 ng/mL final concentration (i.e., on the slide or in the tube).

7. Prior to adding probes or antibodies on slides, we recommend wiping the back and edges of the slide with tissue paper to avoid remaining liquid from "wicking" the small volume of probe/antibody solution off the sections.
8. One of the main causes of disappointing results for any histological protocol is the drying of the samples during the experiment. After fixation, all tissue (whether sections or whole-mounts) should always remain moist. Drying of sections can be avoided by covering with glass or plastic coverslips during long incubation periods and keeping them in humidified chambers. We use plastic boxes at the bottom of which we glue parallel plastic "rails" (e.g., derived from 5 mL plastic transfer pipettes) upon which slides can be placed. Before use, we add wet tissue paper beside and between the rails, so that this does not come into contact with the slides. For whole-mount antennae, ensure the tissue remains submerged in the desired solution in the tubes at all steps.
9. For long-term storage of antennal segments after fixation, wash 3×5 min in PBS+3 % Triton X-100, 3×5 min in PBT and dehydrate 3×5 min in methanol before storing samples in methanol at −20 °C. To rehydrate, wash for 5 min each in 75, 50, and 25 % solutions of methanol in PBT, followed by 4×5 min washes in PBT. Note that RNA may be slightly degraded over time resulting in decreased signal strength, and epitopes of some proteins may be destroyed by methanol. Nevertheless, in most cases we have been able to get good quality stainings from antennae stored in methanol.
10. The antennae can be best seen in very clear polypropylene plastic 1.7 mL tubes (e.g., # MCT-175-C, Axygen, Inc., Union City, CA USA; http://www.axygen.com).
11. The antennae can be mounted in Vectashield directly after the last wash in TNTX buffer; however, we prefer dehydrating and clearing in methanol and BB:BA, as fluorescence signals from these antennae are clearer and sharper. All samples that are mounted in BB:BA need to be fully dehydrated first. Therefore make sure no aqueous buffer remains (e.g., on the lid of the tube), as this will form an emulsion with the BB:BA and make image acquisition very difficult.
12. After mounting, the coverslip will stay in place if it is gently pushed on the nail polish frame. Make sure not to destroy the antennae by applying too much pressure. In our experience, it is best to use 18×18 mm coverslips and about 30 mL of medium. Remove any excess BB:BA with a tissue.

13. To visualize the presence of two different mRNAs in the same sample, double FISH can be performed, with only minor modification to the basic procedure: two differently labeled riboprobes (e.g., Dig-labeled and Fluorescein-labeled) are applied simultaneously. After washing of the unhybridized probe, riboprobe detection is performed sequentially. First, incubate with one of the POD-coupled antibodies (i.e., either anti-Dig or anti-Fluorescein), and visualize with one type of TSA™ Plus system (i.e., Cy5 or Fluorescein) as described above. Quench the POD activity by incubating the samples in 3 % hydrogen peroxide in TNT for 1 h. Remove hydrogen peroxidase by carefully washing (4×5 min washes in TNT) and re-block the samples in TNB. Then, apply the second POD-coupled antibody, and visualize with a different TSA™ Plus system, before final washing and mounting. The length of quenching step may need to be empirically determined, depending upon how much of the first probe is bound. We recommend applying POD-coupled antibodies in the order that will permit detection of the less strongly expressed transcript first.

Acknowledgments

We thank Liliane Abuin, Vincent Croset, Jaime Reina, and Raphael Rytz for comments on this chapter. We acknowledge Elane Fishilevich and Kirsten Vannice (Vosshall Lab, Rockefeller University) and Jeanne de Lavallaz, who contributed to the development of these methods. Research in R.B.'s laboratory is supported by the University of Lausanne, a European Research Council Starting Independent Researcher Grant and the Swiss National Science Foundation.

References

1. Nei M, Niimura Y, Nozawa M (2008) The evolution of animal chemosensory receptor gene repertoires: roles of chance and necessity. Nat Rev Genet 9:951–963
2. Fuss SH, Ray A (2009) Mechanisms of odorant receptor gene choice in *Drosophila* and vertebrates. Mol Cell Neurosci 41:101–112
3. Mombaerts P (2004) Odorant receptor gene choice in olfactory sensory neurons: the one receptor-one neuron hypothesis revisited. Curr Opin Neurobiol 14:31–36
4. Hallem EA, Dahanukar A, Carlson JR (2006) Insect odor and taste receptors. Annu Rev Entomol 51:113–135
5. Touhara K, Vosshall LB (2009) Sensing odorants and pheromones with chemosensory receptors. Annu Rev Physiol 71:307–332
6. Spehr M, Munger SD (2009) Olfactory receptors: G protein-coupled receptors and beyond. J Neurochem 109:1570–1583
7. Hansson BS (2002) A bug's smell/research into insect olfaction. Trends Neurosci 25: 270–274
8. Benton R (2007) Sensitivity and specificity in *Drosophila* pheromone perception. Trends Neurosci 30:512–519
9. Vosshall LB, Stocker RF (2007) Molecular architecture of smell and taste in *Drosophila*. Annu Rev Neurosci 30:505–533
10. Stocker RF (1994) The organization of the chemosensory system in *Drosophila melanogaster*: a review. Cell Tissue Res 275:3–26
11. Nakagawa T, Vosshall LB (2009) Controversy and consensus: noncanonical signaling

mechanisms in the insect olfactory system. Curr Opin Neurobiol 19:284–292

12. Benton R, Vannice KS, Gomez-Diaz C, Vosshall LB (2009) Variant ionotropic glutamate receptors as chemosensory receptors in *Drosophila*. Cell 136:149–162
13. Croset V, Rytz R, Cummins SF, Budd A, Brawand D, Kaessmann H, Gibson TJ, Benton R (2010) Ancient protostome origin of chemosensory ionotropic glutamate receptors and the evolution of insect taste and olfaction. PLoS Genet 6:e1001064
14. Larsson MC, Domingos AI, Jones WD, Chiappe ME, Amrein H, Vosshall LB (2004) *Or83b* encodes a broadly expressed odorant receptor essential for *Drosophila* olfaction. Neuron 43:703–714
15. Benton R, Sachse S, Michnick SW, Vosshall LB (2006) Atypical membrane topology and heteromeric function of *Drosophila* odorant receptors in vivo. PLoS Biol 4:e20
16. Couto A, Alenius M, Dickson BJ (2005) Molecular, anatomical, and functional organization of the *Drosophila* olfactory system. Curr Biol 15:1535–1547
17. Fishilevich E, Vosshall LB (2005) Genetic and functional subdivision of the *Drosophila* antennal lobe. Curr Biol 15:1548–1553
18. Abuin L, Bargeton B, Ulbrich MH, Isacoff EY, Kellenberger S, Benton R (2011) Functional architecture of olfactory ionotropic glutamate receptors. Neuron 69:44–60
19. Dobritsa AA, Goes van Naters W van der, Warr CG, Steinbrecht RA, Carlson JR (2003) Integrating the molecular and cellular basis of odor coding in the Drosophila antenna. Neuron 37:827–841
20. Ai M, Min S, Grosjean Y, Leblanc C, Bell R, Benton R, Suh GS (2010) Acid sensing by the *Drosophila* olfactory system. Nature 468:691–695
21. Grosjean Y, Rytz R, Farine JP, Abuin L, Cortot J, Jefferis GS, Benton R (2011) An olfactory receptor for food-derived odours promotes male courtship in *Drosophila*. Nature 478: 236–240
22. Hallem EA, Carlson JR (2006) Coding of odors by a receptor repertoire. Cell 125: 143–160
23. Silbering AF, Rytz R, Grosjean Y, Abuin L, Ramdya P, Jefferis GS, Benton R (2011) Complementary function and integrated wiring of the evolutionarily distinct *Drosophila* olfactory subsystems. J Neurosci 31: 13357–13375
24. Ramdya P, Benton R (2010) Evolving olfactory systems on the fly. Trends Genet 26: 307–316
25. Hansson BS, Stensmyr MC (2011) Evolution of insect olfaction. Neuron 72:698–711

Chapter 17

Deorphanization of Human Olfactory Receptors by Luciferase and Ca-Imaging Methods

Kaveh A. Baghaei

Abstract

Olfactory receptors (ORs) are expressed at the cell surfaces of olfactory sensory neurons lining the olfactory epithelium and are the first actors for the perception and recognition of odorants.

The olfactory system is mostly activated by ORs and odorants in the cAMP pathway. To date, few human ORs have been deorphanized, mainly because the difficulties encountered in expressing ORs at the cell surface. It has been shown that ORs are able to induce intracellular calcium elevation. Two independent tests, Ca-imaging and CRE-Luciferase assays, have been used in the deorphanization of human ORs and in the analysis of their responses to chemical compounds.

The outcomes of the two techniques typically used for deorphanization have some similarities and differences. The difference in results by CRE-luciferase and Ca-imaging method could reflect technical artifacts such as stimulation or transfection time. There are two main differences in the settings for the two tests: (1) the requested time for odorant exposure and observed response. In the case of Ca-imaging, the increase in fluorescence is recorded between 15 and 30 s, whereas in the cAMP the luminescences are recorded 4–6 h. (2) The time schedule for transfection. Ca-imaging is done 48–72 h after transfection but in cAMP assay it will be done after 24 h. These differences in the setting parameters of the two tests are not the only reasons for variations in results. In vivo analysis has shown that ORs exposed to different odorants can elicit a response via cAMP or IP3 pathway or both.

Our investigations have led us to believe that using the two methods complementarily will benefit OR deorphanization.

Key words Olfactory receptors, Odorants, Deorphanization, CRE-Luciferase, Ca-imaging

1 Introduction

Olfactory receptors (ORs) constitute the largest group of G protein-coupled receptor (GPCR) families. In response to odorants, olfactory sensory neurons (OSNs) convert the chemical odor stimuli into an electrical signal by a cascade of reactions that transmit signals to the brain, resulting in odor perception.

Between 400 and 1,200 different odorant receptors, which are highly polymorphic, are reported in mammals, (1). Several

Chiquito J. Crasto (ed.), *Olfactory Receptors: Methods and Protocols*, Methods in Molecular Biology, vol. 1003, DOI 10.1007/978-1-62703-377-0_17, © Springer Science+Business Media, LLC 2013

mechanisms have been proposed to explain olfactory perception in humans, one of the hypotheses is referred to as the "combinatorial code." Here, different odorants are encoded by different combinations of ORs, but each OR may serve as one component of the unique combinatorial receptor codes for many odorants. Different ORs that recognize the same odorant might recognize different structural features of the odorant. (2, 3).

The binding of odorants to ORs activates the production of cAMP. OSNs express a G protein with a specific subunit named $G_{\alpha olf}$ (4). With the binding of an agonist to the suitable OR, $G_{\alpha olf}$ activates an olfaction-specific adenylate cyclase. The adenylate cyclase activity leads to an increase in cAMP concentration leading to the opening of channels. By permitting the entry of Ca^{2+}, depolarization of the neuronal membrane occurs. This depolarization, amplified by a Ca^{2+} activated Cl^{-} current, generates an action potential. The adenylate cyclase (AC)/cAMP pathway is essential for olfactory responses in vertebrates (5–8). However, odorants activate more than one transduction cascade; indeed, IP3 has also been shown to be an efficient second messenger (9–11). Individual OSNs express adenylate cyclase III and the phospholipase C beta-2, which are needed for the activation of the cAMP and IP3 pathways, respectively (12).

In addition to their importance in detecting odorants, only few OR ligand pairs have been identified and most ORs remain orphan receptors. The basis of odorant recognition by their cognate receptors has been one of the central problems in the study of odor coding. This is due to technical difficulties and the absence of high-throughput methods that connect odorants with OR that they excite.

Several in vitro tests have been developed to deorphanize ORs. Basically, they interrogate the increase of intracellular Ca^{2+} or cAMP resulting from the specific binding of an odorant onto an OR expressed by a transiently transfected cell. It has, however, been observed in vivo that different compounds activate either the cAMP or the IP3 pathways or both (6–11).

2 Materials

2.1 Cell Culture and Transfection

1. Hana3a and HEK 293 are maintained under standard conditions in DMEM supplemented with 10 % FBS, 100 U/ml penicillin and streptomycin, and 2 mM L-glutamine.
2. 2.5 M $CaCl_2$ and 2× HBS for Calcium phosphate transfection, keep frozen at −20 °C.

3. FuGENE transfection reagent from Promega stored at 4 °C.
4. CD 293 medium for transfection.
5. 96-well white plate with Poly-D-Lysine cell ware (Biocoat).

2.2 Ca-Imaging Reagents

1. Ringer's solution for Ca-imaging: 10 mM HEPES (4-(2-hydroxyethyl)-1-piperazineethanesulfonic acid), pH 7.4, 140 mM NaCl, 5 mM KCl, 2 mM $MgCl_2$, 2 mM $CaCl_2$. Store at 4 °C.
2. Fura-2 Acetoxymethylester (Invitrogen) is dissolved at 1 mg/ml in DMSO and stored in single use aliquots at −20 °C (see Note 1).

2.3 CRE-Luciferase

1. Dual-Luciferase® Reporter Assay System kit from Promega including two buffers: (1) Luciferase Assay Reagent II, (2) Stop & GloR Reagent.

According to Promega protocol, Luciferase Assay Reagent II (LAR II) is prepared by resuspending the provided lyophilized Luciferase Assay Substrate in 10 ml of the supplied Luciferase Assay Buffer II. LAR II is stable for 1 month at −20 °C or for 1 year when stored at −70 °C.

Stop & GloR Substrate is supplied at a 50 concentration. Add 1 volume of 50× Stop & GloR Substrate to 50 volumes of Stop & GloR Buffer in a glass or siliconized polypropylene tube. Stop & GloR Reagent may be stored at −20 °C for 15 days with no decrease in activity (see Note 2).

2.4 Immunostaining

1. Hank's Buffered Salt Solution (HBSS), with calcium chloride and magnesium chloride (Gibco).
2. HEPES, 1 M (Gibco).
3. NaN_3 (EM Science).
4. Paraformaldehyde (PFA).
5. Poly-D-Lysine (Sigma).
6. Mouse monoclonal anti-Rhodopsin antibody, 4D2 34.
7. 488Goat Anti mouse (Sigma).
8. Staining solution for first and second antibodies: 5 μl of 1 M HEPES, 5 μl of 1.5 M Natriumazid (NaN_3), 500 μl DMEM + 10 % FBS.
9. Washing solution: 500 ml HBSS, 5 ml of 1 M HEPES, 5 ml of 15 mM Natriumazid (NaN_3).
10. Mowiol 4-88 (Calbiochem).
11. ProLong Gold antifade reagent (Invitrogen).

3 Methods

Two different techniques typically used for the deorphanization of ORs, Ca-imaging and CRE-Luciferase. Here, we used transfected cells by the calcium phosphate method for Ca-imaging and transfected cells by FuGENE reagent for CRE-Luciferase assay. One of the difficulties in ORs deorphanization is the expression of ORs on cell membrane; most of the cells are not able to express ORs on the cell membrane. Live cell surface Immunostaining could be used as a solution to confirm the expression of ORs on cell surface.

3.1 Cell Culture and Transfection

3.1.1 Calcium Phosphate Transfection Method (See Note 3)

1. Cultivation HEK293 cells in Petri dishes (Ø 35 mm) according to standard cell culture protocols.
2. The transfection process could be carried out after 2–3 days, whereby the cell culture was 70–80 % confluent.
3. The calcium-phosphate method, as explained by Loyter (13), is applied to transfect DNA plasmid transiently in HEK293 cells. For each transfection, put 45 μl water in a 1.5 ml sterile tube. Add appropriate ORs (3 μg) with other cofactors in towards the middle of the water and then add 5 μl 2.5 M $CaCl_2$ and 50 μl 2× HBS at the bottom and surface of water in a 1.5 ml tube, respectively. Mix the contents by aeration (see Note 4).
4. Incubate for 15–20 min at room temperature (see Note 5).
5. Disseminate transfection mix to the plates, drop-wise.
6. Incubate the cells for 48–60 h at 37 °C, 5 % CO_2.

3.1.2 Cationic Liposome-Mediated Transfection Methods

1. Aspirate DMEM from a flask that is confluent with HANA cells.
2. Add Trypsin/EDTA (3.5 ml for a 75 cm^2 flask).
3. Incubate at 37 °C for 5 min, until a cell monolayer is detached from flask bottom.
4. Transfer cells to 50 ml conical flask containing 31.5 ml $DMEM^+$ (DMEM supplemented with 1 % Pen/Strep, 5 % FBS).
5. Spin tube at 112 RCF at 4 °C for 5 min to pellet cells.
6. Aspirate $DMEM^+$ being careful not to touch pellet.
7. Resuspend pellet in 1 ml $DMEM^+$, pipet up and down repeatedly to assure total resuspension.
8. Add 19 ml $DMEM^+$.
9. Remove 20 μl of cell resuspension (with 1 ml pipet) and transfer to an Eppendorf tube.
10. Add 20 μl Trypan Blue stain.
11. Count the cells.
12. Plate cells in 96-well plate cover with poly-D-lysine at a density of 3,500 cells/well (in 50 μl volume).

13. Incubate for 24 h at 37 °C in a CO_2 incubator.

Day 2

14. For well one, add the following in two different tubes (scale according to number of wells needed and use the excess to cover loss during pipetting) (see Note 6):

Tube (a)	52 ng receptor plasmid 20.8 ng pCRE-Luci plasmid pSV40-Renilla (10.4 ng) DMEM (no serum) 5 μl per well
Tube (b)	0.15 FuGENE transfection reagent (according to protocol) DMEM (no serum) 5 μl per well

15. After 5 min room temperature incubation, mix solutions **a** and **b** together to create transfection mix.
16. Incubate for 15–20 min at room temperature.
17. Add 45 μl DMEM (with 5 % FBS) to **a** and **b** mixture (see Note 7).
18. Aspirate media from cells in the 96-well plates. Pay attention so as to not disturb the cells. Add 50 μl of DMEM (with 5 % FBS and AB mixture) to each well.
19. Incubate for 24 h at 37 °C in a CO_2 incubator.

3.2 Live Cell Surface Immunostaining (1)

1. Observe the transfected cells under a microscope. The cells in each 35 mm dish should be 50–80 % confluent at the time of staining.
2. Keep cell culture dishes, staining solution and washing solution for immunostaining on ice.
3. Spread the ice into an insulated container to create a horizontal surface. Place a plastic tray with a flat surface on the ice. Cut out a piece of parafilm that is smaller than the tray surface. Spray the inside of the tray with 70 % ethanol and affix the parafilm to the tray to create a smooth surface.
4. Prepare 100 μl primary antibody solution (1/100 mouse anti-Rhodopsin 4D2) for each sample. For example, dilute 1 μl mouse anti-Rho antibody 4D2 in 99 μl staining solution for immunostaining. Store on ice.
5. Transfer the cover glass from the 35 mm dish to the parafilm on the tray, cell-side up. Leave the 35 mm dish with the DMEM medium on ice for later use.
6. Carefully spread 100 μl primary antibody solution onto the cover glass. Make sure that the cover glass is totally covered with the antibody solution. The surface tension of the solution should prevent overflow if carefully applied.
7. Cover the tray. Incubate on ice for 30–60 min.

8. After incubation, transfer the cover glass back to its original 35 mm dish with DMEM. Aspirate the medium from the edge of the dish without touching the cover glass. Gently add 2 ml of washing solution for immunocytochemistry from the edge of the dish. Aspirate. Repeat this washing step two additional times. Do not aspirate after the last wash.
9. Prepare 100 μl of the secondary antibody solution (488 Goat Anti mouse) for each sample. Store on ice.
10. Transfer the cover glass from the 35 mm dish to the parafilm on the tray, cell-side up. Leave the 35 mm dish with wash solution on ice for later use.
11. Spread 100 μl secondary antibody solution onto cover glass so that it covers the glass surface entirely without overflowing.
12. Cover the tray. Incubate on ice for 30 min.
13. After incubation, transfer the cover glass back to its original 35 mm dish with washing solution. Aspirate the solution from the edge of the dish without touching the cover glass. Gently add 2 ml of washing solution for immunocytochemistry from the edge of the dish. Aspirate. Repeat the washing step two additional times.
14. Replace the wash solution from the last wash with 1 % PFA. Cover the tray and incubate for a minimum period of 15 min on ice.
15. Transfer the cover glass from the 35 mm dish and mount it cell-side down onto a microscope slide with a drop of Mowiol mounting solution. Allow Mowiol to air dry. Gently wash the surface of the cover glass and the slide with distilled water.
16. Observe for the OR cell-surface expression under a fluorescent microscope (Fig. 1).

3.3 Ca-Imaging

1. Transfected HEK293 cells (after 48–72 h) were incubated for 30 min in loading buffer (pH 7.4) containing 3 μM Fura-2-AM. Number of the cells to be seeded can be adjusted according to the size of plate/chamber used.
2. Aspirate the medium and remove the extracellular Fura-2 by washing and replacing with Ringer's solution.
3. Prepare chemical odorants (usually 50–500 μM) in Ringer's solution, so that the DMSO concentration does not exceed 1 % in applied solution (stock solution of odorants prepared in DMSO) and injection to column that provided by ratiofluometric Ca-Imaging system.
4. Prepare 3 μM ATP (solvated in Ringer's solution) as the positive control to check the viability of the cells.
5. After each odorant stimulation cells should be washed with Ringer's solution. The final stimulation would be for ATP reaction to check the cell viability.

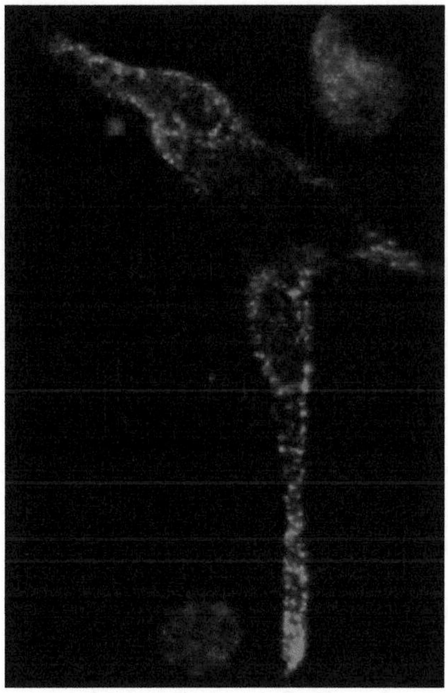

Fig. 1 Live cell surface immunostaining. Expression of olfactory receptors (ORs) on cell membrane is one of the key points in deorphanization procedure. To show and confirm the expression of ORs in cell membrane immunostaining could be used as a solution. Confocal image was obtained on a Zeiss LSM310 microscope using oil immersion lens

6. Odorants and ATP were applied by a specialized microcapillary application system.
7. Using an inverted microscope equipped for ratiometric imaging.
8. WinNT based CellM software served to collect and quantify spatiotemporal Ca^{2+} dependent fluorescence signals (measured as f340/f380 ratio) of HEK cells (Fig. 2).

3.4 CRE-Luciferase

1. Transfected HANA cells on poly-D-lysine-coated 96-well plates are ready for stimulation after 24 h. The cells in each well should be 50–80 % confluent at the time of stimulation.
2. Replace DMEM medium with CD293 (chemically defined medium) and incubate the plate for 30 min at 37 °C and 5 % CO_2.
3. Add 25 μl of odorant solution diluted in CD293 and incubate for 4 h at 37 °C and 5 % CO_2.
4. Incubate the plate with butterfly luciferase buffer (20 μl per well) for 10 min at room temperature.
5. Measuring by plate reader (Luminometer).
6. Add second buffer (renilla/20 μl per well) and incubate for 10 min at room temperature.
7. Measuring by plate reader (Luminometer).

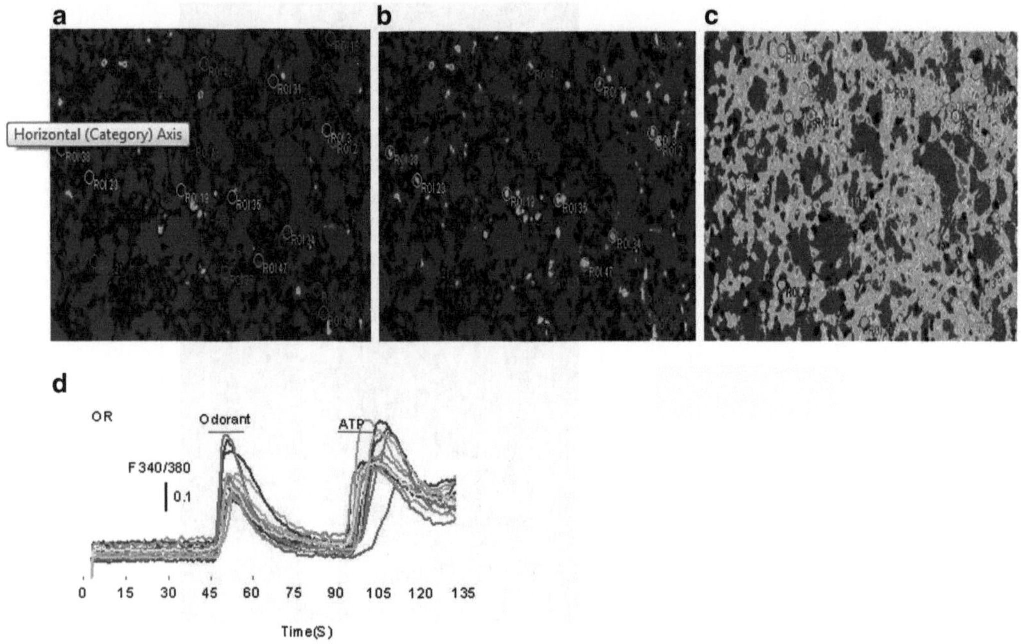

Fig. 2 Ca-imaging. OR activations lead to elevate of the intracellular Ca^{2+} amount. Nonpolar Fura-2/AM is fluorescent but not calcium-sensitive. Once diffuses into cell plasma, Fura-2/AM is hydrolyzed to Fura-2 which is calcium-sensitive and functions as Ca^{2+} chelator with four carboxyl groups. Before stimulation of HEK cells contain ORs with suitable ligand (just with ringer), Fura-2 emits at the wavelength of 510 nm (Fig. 2a). Once Fura-2 binds to free intracellular Ca^{2+} in cell plasma (because of activation the GPCRs cascades), its ratio of emission at wavelengths of 340 and 380 nm is changed and correlated to the amount of intracellular Ca^{2+} (Fig. 2b). According to this feature, the analysis of 340/380 nm excitation ratio for Fura-2 allows to quantify intracellular calcium levels (Fig. 2d). Cells in reaction with ATP (Fig. 2c). Activated cells can be detected and visualized through the analyzing computer system

8. According to manufacturer's protocols for measuring Luciferase and Renilla activities, calculate normalized Luciferase activity with using the formula (Ln − Lmin)/(Lmax − Lmin), where Ln is the luminescence of firefly in response to the odorant, Lmin is the minimum Luciferase value on a plate, and Lmax is the maximum Luciferase value on plate (14) (Fig. 3) (see Note 8).

4 Notes

1. Unless stated otherwise, Fura 2 is absolutely sensitive to light, it should be solvated in DMSO in dark room; it is better to aliquot in every tube, one at a time.
2. For best activity, mix Stop & GloR Reagent exactly before adding to wells.
3. This protocol can be used for HANA cells and other cell lines but the transfection protocol needs to be optimized for different cell lines. The transfection efficiency could be affected by

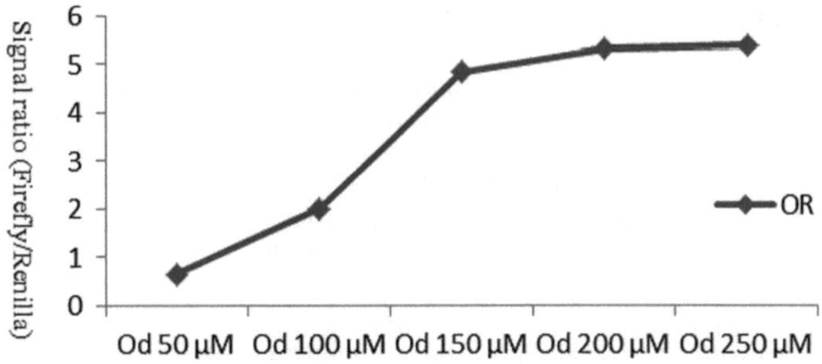

Fig. 3 Dose curve of luciferase assay for odorant (Od) with different concentration. All luminescence values are divided by the Renilla Luciferase activity to control for transfection efficiency in a given well. For data analyzing we used Microsoft Excel

cell type, cell condition, density and plasmid concentration. HANA cells usually are sensitive to high concentration of ORs and cofactors.

4. The sequential process of adding DNA plasmids, $CaCl_2$, and HBS determines the formation of calcium-phosphate crystal. Calcium cations and phosphate anions created crystal precipitations coupled with DNA which could invade the cells through endocytosis. Efficiency of transfection is related to formation of these crystals.
5. 15–20 min is enough time for precipitation of crystals (see Note 4) on the cells. With less than optimal precipitation time, crystals are too small and contain low amounts of DNA. With precipitation time that lasts more than 20 min, the crystals would be larger and not impede their uptake by the cells.
6. Dilute FuGENE reagent by pipetting directly into serum-free medium. For better reaction of FuGENE reagent and to avoid reduction of transfection efficiency, at first diluted the FuGENE reagent in DMEM serum free medium and then add DNA complex top of it. Concentration and purity of DNA are really effective in transfection reaction.
7. Because of serum activities, the cells need to be maintained in the DMEM containing low amount of FBS after transfection. Decrease the FBS will reduce the basal level activity of Luciferase.
8. The normalization of assay results is helpful for comparing data from different experiments.

Acknowledgment

The author would like to thank Professor Hanns Hatt and Dr. Gunter Gisselmann for their advice and encouragement.

References

1. Zhuang H, Matsunami H (2008) Evaluating cell-surface expression and measuring activation of mammalian odorant receptors in heterologous cells. Nat Protoc 39:1402–1413
2. Malinc B, Hirono J, Sato T, Buck L (1999) Combinatorial receptor codes for odors. Cell 96:713–723
3. Gaillard I, Rouquier S, Pin JP, Mollard P, Richard S, Barnabe C, Demaille J, Giorgi D (2002) A single olfactory receptor specifically binds a set of odorant molecules. Eur J Neurosci 15:409–418
4. Belluscio L, Gold GH, Nemes A, Axel R (1998) Mice deficient in G(olf) areanosmic. Neuron 20:69–81
5. Sklar PB, Anholt RR, Snyder SH (1986) The odorant-sensitive adenylate cyclase of olfactory receptor cells. Differential stimulation by distinct classes of odorants. J Biol Chem 261: 15538–15543
6. Brunet LJ, Gold GH, Ngai J (1996) General anosmia caused by a targeted disruption of the mouse olfactory cyclic nucleotide-gated cation channel. Neuron 17:681–693
7. Nakamura T (2000) Cellular and molecular constituents of olfactory sensation in vertebrates. Comp Biochem Physiol A Mol Integr Physiol 126:17–32
8. Wong ST, Trinh K, Hacker B, Chan GC, Lowe G, Gaggar A, Xia Z, Gold GH, Storm DR (2000) Disruption of the type III adenylyl cyclase gene leads to peripheral and behavioral anosmia in transgenic mice. Neuron 27: 487–497
9. Restrepo D, Miyamoto T, Bryant BP, Teeter JH (1990) Odour stimuli trigger influx of calcium into olfactory neurons of the channel catfish. Science 249:1166–1168
10. Bruch RC (1996) Phosphoinositide second messengers in olfaction. Comp Biochem Physiol B Biochem Mol Biol 113:451–459
11. Spehr M, Gisselmann G, Poplawski A, Riffell JA, Wetzel CH, Zimmer RK, Hatt H (2003) Identification of a testicular odorant receptor mediating human sperm chemotaxis. Science 299:2054–2058
12. Noe' J, Breer H (1998) Functional and molecular characterization of individual olfactory neurons. J Neurochem 71:2286–2293
13. Loyter A, Scangos G, Juricek D, Keene D, Ruddle FH (1982) Mechanisms of DNA entry into mammalian cells. II. Phagocytosis of calcium phosphate DNA co-precipitate visualized by electron microscopy. Exp Cell Res 139: 223–234
14. Saito H, Kubota M, Roberts RW, Chi Q, Matsunami H (2004) RTP family members induce functional expression of mammalian odorant receptors. Cell 119:679–691

Part V

Olfactory Receptor Evolution

Chapter 18

Evolution of Olfactory Receptors

Kara C. Hoover

Abstract

Olfactory receptors are a specialized set of receptor cells responsible for the detection of odors. These cells are G protein-coupled receptors and expressed in the cell membranes of olfactory sensory neurons. Once a cell is activated by a ligand, it initiates a signal transduction cascade that produces a nerve impulse to the brain where odor perception is processed. Vertebrate olfactory evolution is characterized by birth-and-death events, a special case of the stochastic continuous time Markov process. Vertebrate fish have three general types of receptor cells (two dedicated to pheromones). Terrestrial animals have different epithelial biology due to the specialized adaptation to detecting airborne odors. Two general classes of olfactory receptor gene reflect the vertebrate marine heritage (Class I) and the derived amphibian, reptile, and mammal terrestrial heritage (Class II). While we know much about olfactory receptor cells, there are still areas where our knowledge is insufficient, such as intra-individual diversity throughout the life time, epigenetic processes acting on olfactory receptors, and association of ligands to specific cells.

Key words Olfactory receptor, Olfactory sensory neuron, G protein, G protein-coupled receptor, Seven transmembrane receptor, Ligand, Odor detection

1 Anatomy of Olfactory Receptors

Receptors are protein molecules lying in the plasma membrane or cytoplasm of a cell and have the purpose of binding signaling molecules, known as ligands. Olfactory receptors are responsible mostly for detecting odor molecules and are expressed in the cell membranes of olfactory receptor neurons. At one end of the neuron is an axon that projects to the olfactory bulb of the brain, the first relay in a system that runs to various brain centers. At the other end are dendrites that extend to the mucosal surface and conclude in various specialized cilia. These cilia are the surface for odor interactions in the cell after ligand binding.

Olfactory receptors are members of the receptor superfamily of G protein-coupled receptors. Only found in eukaryotes, these G protein-coupled receptors are transmembrane proteins that have

Chiquito J. Crasto (ed.), *Olfactory Receptors: Methods and Protocols*, Methods in Molecular Biology, vol. 1003, DOI 10.1007/978-1-62703-377-0_18, the ability to interact with G proteins in order to initiate a signaling cascade that transmits an action potential (1). Martin Rodbell and Alfred Gilman received the Nobel Prize in 1994 for their discovery of G-proteins (2). G proteins function as molecular switches: binding to three phosphate groups (guanosine triphosphate or GTP) activates them and binding to two phosphate groups (guanosine diphosphate, or GDP) deactivates them. There are two large families of olfactory receptors: large G proteins (heterotrimers) and small G proteins (3). Olfactory receptor G proteins are heterotrimeric. One G protein, G_{olf} is largely homologous to the common G protein (G_s), which is found throughout the tissues of the body. G_{olf}, however, has evolved specifically to interact with olfactory neurons (4).

G protein-coupled receptors contain a region called the seven transmembrane α-helix (hence an alternative name for the superfamily of receptors, seven transmembrane proteins). This seven transmembrane α-helix is where biochemical communication between extracellular and cytoplasmic domains takes place. The α-helix region contains seven subproteins with the ability to fold (conform) independently of each other while maintaining the capacity to operate collectively. The extracellular component of the G protein-coupled receptor begins with the first helix (or the N terminal) at the N-terminus and contains three loops, between the second and third, fourth and fifth, and sixth and seventh helices. The intracellular or cytoplasmic component of the G protein-coupled receptor ends with the C-terminus of the seventh helix (the C terminal) and contains three loops beginning with the first helix/N-terminal to the second helix, the third and fourth, and fifth and sixth. G protein-coupled receptor ligands (such as olfactory ligands) bind to the transmembrane domain but some bind to the extracellular domain (1, 5).

There are three major families of G protein-coupled receptors (Table 1) (6, 7). Most have known ligands, but olfactory receptors have few known ligands. This may be due to the complexity of the olfactory receptor system where receptors and neurons are not as specialized as other receptor systems. Family A is the largest and consists of over 900 genes in humans and contains the olfactory receptor subfamily. Olfactory genes constitute the largest number of genes in all the G protein-coupled receptor families.

2 Physiology of Olfactory Receptor Cells

Olfactory receptors do not have ligand specificity to the degree that other receptors do. Instead, each one attracts a range of ligands. Likewise, odorants/ligands may bind to more than one olfactory receptor, but with different degrees of affinity (7). Ligand binding to olfactory receptors occurs in the transmembrane region and activates several intracellular processes. Ultimately, the activated

Table 1
G protein-coupled receptor families

Family	Receptors	Genes
A	Rhodopsin (light), andrenergic (adrenaline), olfactory, and most 7TM types	≈670
B	Gastrointestinal peptide hormone family, corticotrophin-releasing hormone, calcitonin, and parathyroid hormone	>35
C	Metabotropic glutamate receptor family, GABA receptor, calcium-sensing receptor, some taste receptors	>35
D–F	Others	>100

olfactory receptor is the start of a signal transduction cascade that ultimately results in a nerve impulse transmitted to the brain.

The signal transduction of G protein-coupled receptor activation is not clearly understood but can be explained by one of two processes: activation is caused by a conformational change in the receptor after ligand binding, or ligand binding shifts the receptor conformational equilibrium (that exists between active and inactive states) towards the active state. The G protein bound to the G protein-coupled receptor is then activated and detaches allowing the receptor to activate another G protein. Activation may also cause the formation of heterotrimers (comprising the three subunits of the G protein: α, β, γ) or may already be bound to heterotrimers in an inactive state (e.g., bound to GDP). This domain of the receptor is the guanine nucleotide exchange factor, allowing exchanges of bound GDP ("off switch") with GTP ("on switch") on the G_α subunit of the G protein (5, 6).

The role of the G protein is to activate adenylyl cyclase, a lyase enzyme used in catalyzing the breakdown of chemical bonds. Adenylyl cyclase catalyzes conversion of adenosine-5′-triphosphate (ATP, or the nucleotide responsible for transporting energy within the cell) to cyclic adenosine monophosphate, or cAMP. The cAMP nucleotide is responsible for opening ionic channels in the plasma membrane. Once open, the ion channels depolarize the olfactory sensory neuron. Depolarization spreads to the dendrite and soma of the olfactory sensory neuron which triggers action potentials. Action potentials are transmitted to the olfactory bulb via the olfactory neuronal axon. Olfactory neuronal axons converge in the olfactory bulb where initial signal organization and processing take place. Information is then transmitted to the olfactory cortex of the cerebrum where odor perception takes place. Interestingly, olfactory sensory neurons degenerate and regenerate every few weeks, rendering cyclic reconstitutions of the olfactory system (8). Given the regular reconstitution and unknown environmental effect on a potentially plastic system, intra-individual variation in epithelial receptor diversity over multiple reconstitutions throughout a lifetime is unknown.

3 Diversity

There are as many as 1,000 olfactory genes in the mammalian genome. A reduction in functional genes occurred in the primate lineage around the derivation of trichromatic color vision. Strong selection pressure on olfactory receptor genes may have been reduced as a result of the increased reliance on trichromatic color vision. Of all the primate super families, apes are the most derived when it comes to the trend of reduced olfactory receptor gene functionality. The current estimate of olfactory receptor genes in humans is 851 (387 functional genes and 464 pseudogenes, genes that have accumulated enough mutations to render them nonfunctional) (Table 2) (9, 10). While humans have a reduced set of functional olfactory receptor genes, they possess ten million olfactory neurons.

Some research suggests that some olfactory receptors in the pseudogene category still have a weak expression (11). Indeed, a recent study finds that OR pseudogenes may be expressed (transcribed into RNA, but not translated into proteins) in the olfactory epithelium and may share neurons with active olfactory receptor genes (12). Researchers in this area have suggested that interpopulation variation is the result of differential selection events acting on human populations (11, 12).

The array of olfactory receptors unique to a species provides an olfactory repertoire, a biochemical system that discriminates the largest number of odors possible for that species evolutionary ecology. The olfactory receptor system is analogous to the immune system in its diversity and dynamic capacity to identify novel molecules much as the immune system identifies novel pathogens. Humans, despite containing fewer active genes compared to other

Table 2
Distribution of olfactory receptor genes and pseudogenes across the human chromosomes

Chromosome	1	2	3	4	5	6	7	8	9	10	11	12	13	14	15	16	17	18	19	20	21	22	X	Y	Total
Gene	63	2	10	0	4	16	15	1	25	1	167	17	0	23	5	2	14	0	20	0	0	1	1	0	387
Pseudogene	32	9	21	12	4	21	21	11	14	7	202	24	8	22	15	1	4	2	21	0	3	0	10	0	464

mammals, have a vastly diverse olfactory receptor capacity: 178 different functional genomes were found in 189 individuals in a study attempting to quantify this diversity. Only the human leukocyte antigen system contains a similar interindividual variation in the human genome (13). Comparisons between the two systems end there, as olfactory receptor diversity is driven by a large number of genes rather than in situ recombination. Because each odor activates more than one type of olfactory receptor, the possible combinations and permutations are limitless. Thus, the diversity and lack of specialization allow detection of and discrimination among an infinite number of odors.

Diversity of olfactory repertoires—the array of odors detectable and perceivable by an organism–is the result of differential adaptation to environmental challenges. Each species has the capacity to detect and discriminate among a range of chemical signals specific to its own evolutionary ecology that aid an organism in seeking mates, finding food, and avoiding predators. Interestingly, humans show substantial variability in odor perception (13–16) and cultural variation in olfactory hedonic value (13, 14, 17–19). While this variation may be the result of bottlenecks and geographic isolation, selection and cultural practices may also be active in conserving olfactory receptor genes (11). This is another area of research that has yet to be investigated in a systematic manner.

4 Evolution of Olfactory Receptors

The olfactory receptor gene family in vertebrates evolved by birth-and-death evolution. Birth-and-death evolution is a special case of the continuous-time Markov process, a stochastic process that describes a situation in which the evolutionary transition from one state to the next is indeterminate based on the existing parameters. This is captured via equations that describe the probability of each possible pathway to the next state. Birth-and-death evolution is a special case of this continuous-time Markov process. In birth-and-death evolution, each state is described by the size of a population and each transition is limited by births and deaths. In the case of the olfactory sub-genome, the creation of new genes occurs through gene duplications (births), gene conversions (deaths and births), and other genomic events (20, 21).

The evolution of vertebrate olfactory receptor genes follows the expected pattern of three separate evolutionary lineages: fishes, amphibians, and mammals (22). Fishes retain eight out of nine ancestral olfactory receptor gene lineages. Despite retaining the largest number of olfactory receptor gene lineages, they have the smallest number of olfactory receptor genes (≈100). Within this small number of genes, we find the greatest amount of variation in all major classes of animals. Retention of these artifacts of both genetic structures and genes as well as the diversity of genes reflects a longer evolutionary history in fish compared to any other vertebrate (23, 24). Birds and mammals retain the smallest number of gene lineages (two) but have the greatest number of genes (≈1,000). Indeed, olfactory receptor genes are the largest mammalian gene family. Amphibians retain gene families common to both fishes and mammals reflecting their status as a transitional vertebrate capturing both volatile and waterborne odorants; however, amphibians share more genes with mammals than fish, perhaps a reflection of their life history that begins in the water but the majority of which is lived on the land (25). Roughly 200 million years ago an olfactory receptor gene radiation occurred, coinciding with the time of reptile dominance (24). This genetic radiation suggests a major adaptive shift and/or speciation event at this time, likely related to the terrestrial environment.

Phylogenetic analysis of the vertebrate olfactory receptor family suggests that there are two main types of olfactory receptor genes distinct early on in vertebrate history: Class I and Class II (26). Specialization of these olfactory receptor genes to a particular medium, air or water, is implied by the respective association of Class I genes with fishes and water-soluble odorants and Class II genes with mammals and airborne (volatile) odorants (27, 28). Indeed, five Class I and one Class II genes are common to both fishes and tetrapods (22). Unfortunately, mammals also have active Class I genes (even humans retain 100), making this distinction murkier (25, 29).

As an evolutionarily older class of animals, fishes have primitive or ancestral olfactory receptor genes yet they exhibit tremendous variation among individuals and species (22, 26). The long evolutionary lifespan of fishes also has resulted in tremendous diversity in olfactory organs (30). Fishes smell by moving water through the nares which bypass the pharynx and lead straight to the olfactory rosette which is lined with sensory epithelium. This epithelium contains three types of sensory cells: one for odorants and two for social cues and sex pheromones (31). An order of fishes, coelacanths, is thought to represent a transitional state between fishes and tetrapods due to the fact that they possess leg-like structures. Known originally from fossils, there are two living species left (32). One species, *Latimeria chalumnae*, is particularly relevant to a discussion on the evolution of olfaction. *L. chalumnae* share a class of olfactory receptor genes with amphibians and mammals, but not other fishes (25).

Early tetrapods and modern amphibians have the capacity (at different life stages) to detect both waterborne and airborne odors. This dual adaptation allowed early tetrapods (and modern amphibians) to reap the benefits of resources in both environments. Their genes and biology reflect a dual specialization for detecting both volatiles (airborne) and water-soluble odorants. In the water, amphibians detect odorants in a manner similar to fishes—flooding of the nasal cavity with water via the respiratory pump (33).

As discussed earlier, Class I olfactory receptor genes are commonly associated with water-soluble odorant detection and Class II olfactory receptor genes with volatile odorant detection. Indeed, Class II olfactory receptor genes in both *L. chalumnae* and aquatic mammals are nonfunctional (pseudogenes). A secondary class of olfactory receptor genes evolved in early proto-tetrapods and underwent selective pressure in tetrapods. These olfactory genes were pseudogenized in mammals that returned to the sea. What was the original function of these genes in coelacanths that lived in the water? Were they under selective pressure in the coelacanths for some reason or did these Class II olfactory receptor genes assume a new function in terrestrial vertebrate olfaction (29)? The frequency of duplication and gene loss in the vertebrate evolutionary history of olfactory receptor genes suggests adaptive processes playing a strong role in functional variation across these divergent lineages (26).

Mammalian olfactory genetics are divergent from the vertebrate lineage in the sheer size of the olfactory receptor gene family. This gene family comprises 1 % of the functional genome and is the largest gene family identified among species (34). This expansion may reflect mammalian olfactory adaptive radiation (secondary to that of the reptilian olfactory radiation 200 million years ago). Mammalian olfactory receptor genes are located on most chromosomes with a tendency to cluster near chromosome ends, or telomeres (24). Telomeres serve as chromosomal caps that conserve the integrity of DNA during cell division. Recombination (or reshuffling of homologous chromosomal DNA) occurs frequently toward the telomeres. One result of recombination is gene duplication. Telomeric loci for olfactory genes may reflect previous olfactory receptor gene expansion in evolutionary history wherein individual genes or genomic segments were duplicated rapidly (24). In mammals, the overall biological olfactory apparati and genes are conserved—even in those species that returned to water. The largest difference between mammals and fish is the lack of an organ to detect pheromones. While many argue that the lack of the vomeronasal organ in humans renders us incapable to detecting pheromones, we have functional genes in the olfactory epithelium for pheromone detection (35) and individuals without a vomeronasal organ are able to detect pheromones (36).

While various researchers in different branches of the sciences and social sciences have uncovered a tremendous wealth of knowledge about the olfactory system, there are still many questions that remain open to investigation. These include several topics, a few of which are mentioned here: the variation among human populations in receptors, changes in receptor variation over several epithelial reconstitutions, the complete signaling cascade pathway from ligand binding to neural impulse, olfactory ligand–receptor affinity, the role of pheromone receptors in humans, evolutionary causes of birth and death events shaping primate olfactory genes, and the functional relationship in humans between the small number of receptors and huge number of neurons.

References

1. King N, Hittinger CT, Carroll SB (2003) Evolution of key signaling and adhesion protein families predates animal origins. Science. doi:10.1126/science.1083853
2. Marx, J. (1994) Nobel prizes. Medicine: A signal award for discovering G proteins. Science. doi: 10.1126/science.7939678
3. Neves SR, Prahlad RT, Iyengar R (2002) G protein pathways. Science. doi:1s0.1126/science.1071550
4. Jones DT, Reed RR (1989) G_{olf}: an olfactory neuron specific G protein involved in odorant signal transduction. Science. doi:10.1126/science.2499043
5. King, N., C.T. Hittinger and S.B. Carroll (2003) Evolution of key signaling and adhesion protein families predates animal origins. Science doi: 10.1126/science.1083853
6. Pierce KL, Premont RT, Lefkowitz RJ (2002) Seven-transmembrane receptors. Nat Rev Mol Cell Biol. doi:10.1038/nrm908
7. Fredriksson R, Lagerstrom MC, Lundin LG, Schioth HB (2003) The g-protein-coupled receptors in the human genome form five main families. Phylogenetic analysis, paralogon groups, and fingerprints. Mol Pharmacol. doi:10.1124/mol.63.6.1256
8. Menini A, Lagostena L, Boccaccio A (2004) Olfaction: from odorant molecules to the olfactory cortex. News Physiol Sci. doi:10.1152/nips.1507.2003
9. Olender T, Feldmesser E, Atarot T, Eisenstein M, Lancet D (2004) The olfactory receptor universe—from whole genome analysis to structure and evolution. Genet Mol Res 3(4):545–553
10. Safran M, Chalifa-Caspi V, Shmueli O, Olender T, Lapidot M, Rosen N, Shmoish M, Peter Y, Glusman G, Feldmesser E, Adato A, Peter I, Khen M, Atarot T, Groner T, Lancet D (2003) Human gene-centric databases at the Weizmann Institute of Science: GeneCards, UDB, CroW 21 and HORDE. Nucleic Acids Res. doi:10.1093/nar/gkg050
11. Menashe I, Lancet D (2006) Variations in the human olfactory receptor pathway. Cell Mol Life Sci. doi:10.1007/s00018-006-6111-x
12. Zhang X, De la Cruz O et al (2007) Characterizing the expression of the human olfactory receptor gene family using a novel DNA microarray. Genome Biol. doi:10.1186/gb-2007-8-5-r86
13. Menashe I, Man O et al (2003) Different noses for different people. Nat Genet. doi:10.1038/ng1160
14. Gilbert AN, Wysocki CJ (1987) The Smell Survey Results. Natl Geogr 122:514–525
15. Keller A, Zhuang H et al (2007) Genetic variation in a human odorant receptor alters odour perception. Nature. doi:10.1038/nature06162
16. Lancet D, Sadovsky E et al (1993) Probability model for molecular recognition in biological receptor repertoires—significance to the olfactory system. Proc Natl Acad Sci USA 90(8):3715–3719
17. Classen C (1994) Aroma: the cultural history of smell. Routledge, Oxford
18. Lancet D, Benarie N et al (1993) Olfactory receptors—transduction, diversity, human psychophysics and genome analysis. Ciba Found Symp 179:131–146
19. Pandya V (1993) Above the forest: a study of andamanese ethnoanemology, cosmology, and the power of ritual. Oxford University Press, Oxford
20. Nei M, Rooney AP (2005) Concerted and birth-and-death evolution of multigene families. Annu Rev Genet. doi:10.1146/annurev.genet.39.073003.112240
21. Nowak MA (2006) Evolutionary dynamics: exploring the equations of life. Harvard University Press, Cambridge
22. Niimura Y, Nei M (2003) Evolution of olfactory receptor genes in the human genome. Proc Natl Acad Sci USA. doi:10.1073/pnas.1635157100
23. Glusman G, Yanai I et al (2001) The complete human olfactory subgenome. Genome. doi:10.1101/gr.171001
24. Fuchs T, Glusman G, Horn-Saban S, Lancet D, Pilpel Y (2001) The human olfactory subgenome:

from sequence to structure and evolution. Hum Genet. doi:10.1007/s004390000436

25. Freitag J, Ludwig G, Andreini I, Rössler P, Breer H (1998) Olfactory receptors in aquatic and terrestrial vertebrates. J Comp Physiol A. doi:10.1007/s003590050287
26. Shi P, Zhang J (2009) Extraordinary diversity of chemoreceptor gene repertoires. In: Meyerhof W, Korsching S (eds) Chemosensory systems in mammals, fish, and insects. Springer, Berlin
27. Niimura Y, Nei M (2005) Evolutionary dynamics of olfactory receptor genes in fishes and tetrapods. Proc Natl Acad Sci USA. doi:10.1073/pnas.0501922102
28. Niimura Y, Nei M (2006) Evolutionary dynamics of olfactory and other chemosensory receptor genes in vertebrates. J Hum Genet. doi:10.1007/s10038-006-0391-8
29. Collin SP (2007) Nervous and sensory systems. In: McKenzie DJ, Farrell AP, Brauner CJ (eds) Primitive fishes. Elsevier, London, pp 121–179
30. Hara TJ (1975) Olfaction in fish. Prog Neurobiol 5:271–335
31. Hamdani EH, Døving KB (2007) The functional organization of the fish olfactory system. Prog Neurobiol. doi:10.1016/j.pneurobio.2007.02.007
32. Janvier P (2007) Living primitive fishes and fishes from deep time. In: McKenzie DJ, Farrell AP, Brauner CJ (eds) Primitive fishes. Elsevier, London, pp 1–53
33. Reiss JO, Eisthen HL (2008) Comparative anatomy and physiology of chemical senses in amphibians. In: Thewissen H, Nummela S (eds) Sensory evolution on the threshold: adaptations in secondarily aquatic vertebrates. University of California Press, Berkeley, pp 43–63
34. Buck LB (2000) The molecular architecture of odor and pheromone sensing in mammals. Cell. doi:10.1016/j.cub.2006.08.030
35. Wysocki C, Preti G (2004) Facts, fallacies, fears, and frustrations with human pheromones. Anat Rec. doi:10.1002/ar.a.20125
36. Frasnelli J, Lundström JN, Boyle JA, Katsarkas A, Jones-Gotman M (2010) The vomeronasal organ is not involved in the perception of endogenous odors. Hum Brain Mapp. doi:10.1002/hbm.21035

INDEX

Chiquito J. Crasto (ed.), *Olfactory Receptors: Methods and Protocols*, Methods in Molecular Biology, vol. 1003, DOI 10.1007/978-1-62703-377-0, © Springer Science+Business Media, LLC 2013

H

I

T

U

V

W

X

Y

Z